D0170221

Outwitting Contractors

Outwitting Contractors

BILL ADLER, JR.

THE LYONS PRESS

To Karen Robin Adler, who inspired our renovation
because we knew she would be here soon.

Contents

Acknowledgments

FRANKLY, THIS BOOK was hard work to write, but it was fun. I never could have done it alone, however. My greatest thanks to Paul Locher, president of Locher Design-Build, whose technical advice and stories were invaluable. Paul was also the general contractor for the renovation on our eighty-five-year-old house, and I promised him that if the renovation were a success, I would give his company a plug on this page. Here it is: Locher Design-Build, 4509 Brandywine Street, N.W., Washington, D.C. 20016; (202) 966-4448. He did a great job on these pages and on our house.

My wife, Peggy, not only gave me moral supposrt while writing this book, but offered fine editorial suggestions. "Editorial suggestions" is a nice way of saying that she pointed out many stupid things I wrote in the early drafts of OUTWITTING CONTRACTORS. As always, I could not have completed—or even have started—this book without her.

Thanks, too, to Karen Sagsetter and Bruce Hathaway for their sage wisdom about remodeling.

Jane Dystel, New York's ace literary agent, was a source of inspiration and great help. I think it's fair to say that she enabled the idea for this book to become a reality.

Introduction

REMODELING A HOME or apartment isn't a job—it's a war. In 1990 North Americans spent more than $125 billion on home remodeling. That's more than the defense budget of many countries. Each year tens of millions of people add bathrooms, build additions, add fireplaces, enlarge bedrooms, turn driveways into garages, turn garages into dens, reroof, relandscape, tile kitchens, replace floors, blow in insulation, finish basements, create pantries, add family rooms, and build houses from scratch. Some homeowners do the work themselves; the majority hire contractors or laborers.

To make these miraculous changes, we must confront an army of workmen whose purpose, on paper, is to make our home a more enjoyable place but who, in reality, are there to do battle with us at every step along the way.

I wish I could say renovation is going to be fun. It is going to be everything but fun. Even if you're not living through it, even if you're remodeling before you move into your new house, renovation is going to be unfun.

There are plenty of books on how to remodel a house or apartment. There's no lack of information about selecting kitchen tiles, sanding floors, shingling roofs, wiring ceiling fans, and so forth. But no book—until now—has addressed the true needs of most home remodelers: how do you defend yourself against the hordes of invaders who come into your home, renovate according to their remodeling desires (not according to your architectural plan that cost $2,000); destroy personal possessions; decide that there's nothing wrong with a hot water pipe through the middle of the living room; leave foreign objects in the refrigerator; take whatever isn't tied down; treat your telephone like a WATS line; arrive unexpectedly at 6:00 A.M. on Monday, ask you to let them in at the same time on Tuesday, but don't show again up until Thursday; never lock your door; let the dog out to roam; sand the floor while the walls are still wet with paint; paint a room white and use the freshly painted wall as a napkin; ignore the three doormats you conscientiously placed outside and track mud onto oak floors; install whatever fixtures they happen to have on hand from a previous job; use your house like a tobacco-products test center and consider all open spaces ashtrays; listen to their favorite radio stations while you're sleeping; use your bedroom as storage space for bricks, floorboards, and bathroom fixtures; order bathroom supplies that take three months to arrive—just as the plumber goes on vacation; imprint deep, lasting scratches on anything new, valuable, or prominently placed; go bankrupt just as you're prepared to sue, and don't speak English.

These people are important. You can't remodel without them. Fortunately, many of these plumbers, electricians, demolishers, and other house specialists understand their jobs well. But let's be frank: On the whole, remodeling crews aren't known for their artistic sense or delicate touch.

It's you versus the workers.

While this book isn't going to make renovation enjoyable, it is going to make your house better and make your life during renovation a little safer and saner. The concept behind OUTWITTING CONTRACTORS is that there's a wealth of information out there about doing a superior renovation, and the information comes from other people's mistakes. Anybody who has remodeled a home or apartment knows how to do it better next time.

This book doesn't assume that you will get dirty and do your own carpentry, excavating, wiring, or anything at all. Just the opposite. OUTWITTING CONTRACTORS assumes that you want to keep your fingernails clean. But you will learn a lot about how houses and apartments are made and what holds them together. This isn't a technical book, but I do describe the kind of work that subcontractors perform—or, more accurately, are supposed to perform—on your house. My philosophy is that the more you know about what workers should be doing, the less chance there will be of something going wrong. Once you've finished reading OUTWITTING CONTRACTORS you'll sound like an expert; you'll never be intimidated by technical-talking contractors and subcontractors; and you'll never be cheated by them.

There are many analogies you can make between home renovation and other activities. Renovation and medicine come to mind: A surgeon knows how to save your life; a tradesman knows how to save your house. The only difference—a major one, however—is that the surgeon may have decades of experience, while the carpenter's apprentice has been on the job a couple of weeks.

Most of us never learned about renovation in school. There is a long list of renovation skills that most of us simply have no experience whatsoever with: designing plans that work for the way we live, soliciting bids, judging construction, writing construction contracts, determining cost of construction, supervising or managing tradespeople, knowing which materials should be used where, locating and buying supplies, knowing the difference between cost-saving and life-threatening building shortcuts, knowing in which order work should be done, inspecting work, and determining if work is done right. Construction seems to be old—boring—technology. But in fact, construction is a complicated and exacting service with the potential to work wonders or create disasters.

Renovation is a service, not a product. You are hiring individual people, not buying a product. No two houses, apartments, general contractors, payment schedules, and construction teams are identical. Even if you have renovated before, you should not expect your second renovation to be anything like the previous one. Your first renovation, a kitchen, may have been fine, but even with the same company two years later, your second renovation, a bathroom, can be an unmitigated disaster simply because the foreman was a different person.

OUTWITTING CONTRACTORS offers a fresh perspective and valuable information about home remodeling. The book tells you how to deal with architects, contractors, builders, laborers, demolition crews, county inspectors, and sobbing spouses with openness and humor. Although home renovation is a serious (and expensive!) subject, unless you manage your home or apartment remodeling with some humor, you'll simply go crazy.

And crazy isn't too far from the literal truth. According to one prominent psychologist, remodeling is one of the most stressful things a person can do. Some marriages don't survive the process. *The Los Angeles Times* wasn't exaggerating when it gave this headline to a 1988 article* about remodeling: "Altered Estates: Fixing up the House May Wreck the Psyche."

What is renovation? For the purposes of this book, renovation or remodeling is any change in your house that involves outsiders working in your house. A renovation can be as simple as wallpapering a room or as complicated as building a 4,000-square-foot addition. It can be a well-thought-out addition or an emergency roof repair. Whatever kind of renovation you do, this book will make you well armed.

* *Los Angeles Times* (July 28, 1988): part 5, page 12.

A microwave oven is extremely important.

Craig Stoltz, renovation survivor

1 | Surviving Your Renovation

YOU CALL THIS LIVING?
Just about the only thing worse than living through home renovation is . . . well, I guess there isn't anything.

The first step you should take is to schedule your renovation
for when nothing stressful is going on in your life. Simultaneously trying to have a baby and renovating (even if it's to create
a baby's room) is not a good idea. Privacy is something you don't
get a whole lot of while you're living through renovation. Sleep
becomes a scarce resource too. During the few hours when
there's no work going on you'll be clearing dust off your tooth-
brush and bed, trying to find a functioning light switch ("Didn't
they rewire yet?!"), vacuuming, yelling at the contractor* over
the phone, taking throat lozenges, getting calls from subcontractors' ex-wives at 3:00 A.M., and planning a weekend away, then
canceling it because you're afraid to leave your house alone with
those maniacs.

Looking for a new job and conducting a renovation don't go
hand in hand, and this isn't the time to buy new clothes, either.

If this brief description of renovation sounds farfetched to you,
just wait until the guys with the sledgehammers arrive.

*Throughout this book the terms *contractor, general contractor, G.C., foreman,* and
builder are used synonymously.

1

2

One family in Van Nuys, California, coped with their renovation this way: They lived in a motor home in their driveway for weeks. To use a bathroom they had to climb into their house through a window.*

Despite what your friends tell you, living through home renovation is not like a combination of living in Beirut, over a motorcycle bar, in Saudi Arabia during a dust storm, and adjacent to the New Jersey Turnpike. It is worse.

The best way to live through a renovation is not to live through it. Move out. If you can, remove your family and all your worldly possessions from the house. Obviously, moving isn't practical for most homeowners, but it's worthwhile if you can. Don't move too far, because you'll want to inspect the renovation at least once a day. Another reason in favor of vacating the premises is that an empty house makes the remodeling go almost twice as fast and cost less.

Shoes Off, or You've Got to Be Kidding

Expect the worst. My wife and I made it pretty clear to the workers who were installing our closet system at the end of our project that we wanted them to wipe their shoes on the doormat. (Actually we had put two doormats outside just to emphasize the point.) Call us picky, but we didn't want mud on the new wall-to-wall carpeting in the closet. To our surprise, they were obedient, and after their first visit we felt confident that they would be walking in with clean shoes.

It only took a second, but there was nothing I could do when I saw one of the workers diligently and thoroughly wiping his muddy feet on the small, expensive Navajo rug we had just put outside our bedroom. To him it was just another doormat.

You can't expect that workers will treat anything in your house with respect. You can't presume manners of any kind.

If you have a ten-inch diamond necklace lying around and a worker happens to need a ten-inch cord, he might use your necklace. Your crystal bowl looks suspiciously like an ashtray to a

*Barbara Gray, "Long Haul: Three Homeowners Say Trouble by Yard Yields Profit by Mile." *Los Angeles Times,* (July 28, 1988): part 5, p. 12.

worker who's smoking. The fact that floors have been finished recently won't keep workers from dragging an air conditioner across it. A seventeenth-century desk is a perfect place to put a wet metal clipboard ("Sign here, ma'am, where it says that there was no damage done during the work"). A sentimental piece of driftwood that's resting on your fireplace mantel can become a doorstop.

In an instant, a worker can be on his way to damaging something valuable, sentimental, difficult to repair, or on loan from the public library. The nanosecond you see something happening you don't like,

SPEAK UP.

When it's faster to grab a clipboard out of a worker's hand before it leaves a dent on your table, grab first and explain later. You won't insult him, and you certainly shouldn't feel embarrassed about wanting to keep your house scratch, dent, and water mark free. Workers are used to homeowners telling them where they can and cannot walk, but without specific instructions from you, they will do things as they see fit.

IMPORTANT FACT

SIMILARITIES BETWEEN SURGERY AND HOME REMODELING

Surgery on the Body	*Home Remodeling*
There's nothing more important than your body's health.	There's no possession more important than the house you live in.
Surgery is very expensive.	Remodeling is very expensive.
Surgery is a long, involved process requiring much preparation and therapy.	Home remodeling takes a long time and never seems quite done.
A lot can go wrong during surgery; you could become a vegetable or die.	A lot can go wrong during home remodeling; your house could be wrecked or totally destroyed.

4

Surgery is going to make your life miserable for a while.	Home renovation is going to make your life miserable for a while.
You'll gain new insights into your emotional make-up.	You'll discover an aggressive, nasty, impatient self you never knew.
For the rest of your life, you'll have to live with whatever the surgeon does to you.	For the rest of your life, you'll have to live with what *they* do to your house.

THE MAJOR DIFFERENCE BETWEEN SURGERY AND HOME REMODELING

| *Surgery* | *Home Remodeling* |
| Surgery is performed by the surgeon, a skilled, highly educated professional with years of experience and training, as well as devotion to the job. | Home renovation is not performed by the architect, but by people who may or may not have graduated from high school and who really are looking forward to Miller time. |

Pets

Pets don't seem to understand what renovation is all about. Some (most notably cats and goldfish) are terrified of what's going on and may require psychological counseling afterward. Others, including dogs, monkeys, and large parrots, think that home renovation is the greatest entertainment since *The Discovery Channel.* These pets love the excitement, the new toys left lying around, and the freedom of open doors all day long.

As a rule, pets and home renovation don't mix. The renovation is bad for your house and bad for your pets. And in the most unpredictable ways. Michael Weiss and Phyllis Stanger of Washington, D.C., discovered this by watching Roxanne, their terrier, deposit a tennis ball into a yet ungrated floor duct. Says Michael, "Imagine what's going to happen to the tennis ball over the years. I think this is how we'll get Legionnaires' disease."

Strive to separate your pets and your renovation project with as much space as possible.

Dogs love to carry off objects that workers leave around, such as screwdrivers, knives, and faucets. Cats will hide under debris. Birds will acknowledge an open window as an invitation to freedom. Goldfish bowls will quickly come to resemble toxic waste sites.

A pet can do an amazing amount of damage to a home under renovation (even more than the workers). Sometimes this damage is planted like a time bomb for the far future: A cat hides a captured bird in an exposed wall. Sometimes it's immediate: Your dog knocks over new windows before they're installed.

But the damage renovation can do to your pet is even worse. Real psychological traumas can result from exposure to continual loud noise, from surprising loud noises, from disruption of schedules, from strangers (friend or foe?) all around, from strange sights and smells, from surroundings that are undergoing uncomprehensible change. Life with a post-renovation–stressed pet is a challenge. Physical damage can also be significant. Dust and toxic fumes that accompany renovation do no good for your pet's lungs, eyes, skin, and sense of smell. Domestic animals are not designed with built-in protection from this kind of pollution.

While researching this book I heard more than one sad story about cats crawling into small spaces laid open during construction and disappearing. The owners understandably thought that their cats had disappeared into the night through a door left open by a worker. Little did they know that their cats were still with them, turning—over the years—into fossils. If your cat disappears during renovation, don't assume that it has run away. Even if you don't hear any meowing, the cat may have become trapped—and killed—somewhere in the walls or between floors. If your cat becomes "lost," look for it first in the house.

Consider sending your pet to a kennel during the worst renovation days—during demolition, floor refinishing, plastering, and basic carpentry. These are the nights when you also might consider vacating the house. At other times, keep your pet from gaining access to the parts of the house under construction. This may be difficult, but think of it as a test of your creativity.

Sex and Home Renovation

Poor Debbie, a Baltimore housewife. Happily married with two small children, eight months and two years old, Debbie was given the responsibility of watching over the construction of a two-story addition. When IT happened, it was gradual, without anybody expecting it. After the heavy building, plastering, painting, flooring, electrical installation, insulation, and plumbing were completed, the contractor, Steve, still had a substantial amount of post-construction work to do, mostly carpentry. Debbie and Steve spent days together, and then one day, the general contractor followed Debbie to her bedroom for a frolic. Their one-afternoon affair took place while Debbie's children were playing in the yard.

The story has a happy ending, fortunately (something that can't be said for all renovation tales). I'm pleased to report that, after counseling, Debbie's marriage is intact. But, most significant, the contractor also did a good job on the house.

June, who knows Debbie well and suffered through an extensive renovation herself, analyzed Debbie's predicament this way: "When you're stuck at home with small children, you have a sense of impotence—even when there's no renovation going on. It's easy to be unhappy staying at home with children. You have no control over the children, over money, over the mess of the construction. Then, all of sudden you have all these men coming in and you are *in control:* They are doing what you want.

"You spend time with them," June continues, "plus you're under a lot of strain. Your kids are unhappy. Your husband wants to know what's going on with the work. Can you see how Debbie's affair happened? I know this seems really weird—having sex with your contractor—but I bet it's happened more times than you can imagine. Obviously, when you have a crew in your house, you're not going to pick one guy and sleep with him," she concludes. "But this guy was in Debbie's house by himself, day after day."

The mixture of housewives and workers doesn't inevitably lead to sex. Sometimes there's a different consequence. For the first time in years or decades, women who have been working as homemakers, spending eight or nine hours every day for years with only children as company, have plenty of adults to interact with during construction. It's a novel, uplifting experience for people who have been trapped in their homes for the majority of their adult lives. Consequently, some women have a better time coping with renovation than their husbands. The husbands come home each evening, tired from work, maybe cranky, maybe even bummed out about coming home to a giant mess. Their kingdom is disrupted, out of control. Yet the wives are in control all day long. They see progress as it happens: pipes laid, trim installed, fixtures put in place. There may be a general contractor on site, but the housewife is the ultimate boss. Throughout the day she makes dozens of decisions—decisions that will affect her family's life for a long time.

These differences in attitude and relationship to renovation affect the respective ability of husbands and wives to cope with renovation stress. The result? Tension and conflict between spouses.

During any renovation, each partner should watch for warning signs that indicate a dangerous shift in their relationship. These signs include

- One spouse who likes to talk about the renovation (including planning for it) and another who doesn't;
- Yielding decision-making authority to the other spouse; "You decide," said with a hasty, gruff tone;
- Becoming quick tempered;
- An increase in illnesses that have no apparent cause, such as allergies, fatigue, and nausea;
- Listlessness; and
- Not hearing what the other spouse says.

I'm not sure there's a quick way to resolve renovation-induced conflict once it's taken root. Speaking not as a psychologist but as one whose marriage endured renovation, the only practical way to cope with this conflict is to try to prevent it from evolving. Once you notice the symptoms, take action. Therapy may or may not work, but at least it will get you out of the house. Just be prepared: Renovation is very, very stressful. Renovation is like being stuck in a time warp; the future portends wonders, but the present seems to go on forever. Your remodeling will one day be over, and there's no sense in your marriage being over by the time your renovation is done.

Kids

If pets and relationships are a chore to keep happy and healthy during renovation, imagine what kids are like! Especially infants. Or especially five-year-olds. No, make that children around twelve. It really doesn't matter, because children and renovation are about as happy together as animal rights activists and the National Rifle Association.

There are two basic dilemmas when it comes to kids and renovation. All other problems stem from these elemental predicaments. First, there's health: As unhealthy as renovation is for adults and pets—all that dust, those fumes, the noise—it's worse for children. Airborne debris can precipitate allergies, dust can get in eyes, and asbestos can lodge in children's lungs (because it takes such a long time to manifest damage, asbestos, ironically, has its greatest effect on children). Young children like to put everything in their mouths, and older children like to play in dangerous places. Not only do children not know enough to avoid pockets of construction, but they are attracted to hazards. If your house or apartment was once painted with lead paint, renovation will unleash dangerously high concentrations of atmospheric lead, and a child's lungs offer an effective gateway into the bloodstream for air-borne lead.

Second, there's a psychological problem: Young children don't understand the necessity for renovation. From ages one to about four, children see the world through "I"—they are the center of the universe, and when their universe is disturbed, all sorts of unpredictable and bizarre consequences are possible.

Don't renovate when your kids are around.

Take the case of Alice Powers, who lives in a three-story house in Washington, D.C. Alice's two bathrooms were put out of commission for more than three months by a plumber whose skills were, shall we say, not up to NASA standards. Anyway, fortunately for Alice, most of the work took place during the summer, so her three-year-old daughter, who had recently been toilet-trained, was retrained to "go in the yard"—easy to do, since the little girl was also afraid of the "big men" inside. "One day," Alice reports, "we went to visit my brother, Michael, who lives in an apartment, but who has a plot in a community garden across the street. We were in Michael's garden and I said to my daughter, 'Look at Uncle Michael's nice garden.' She dropped her pants and did a big B.M. right there. To her, 'garden' meant 'bathroom'."

Alice discovered why it's a bad idea to renovate with a kid who's around three. They can't understand anything about why all this is happening. Alice's child was intimidated by the men in the house.

If there's any truth to the adage that you learn from your mistakes, Alice Powers is now a brilliant woman. "I have learned that I will never remodel when children are out of school," she says. "They felt robbed of their summer vacation. I constantly had to wait for the plumber and couldn't take my kids out"—a lost summer for her three children, ages three, seven, and ten. "If my kids had been in school they would have had some sense of normalcy. When they look back on that summer, they feel that it was the worst summer they ever had. They were really angry."

But older kids have a little better time coping with renovation. Older children can anticipate: They have something to look forward to, often a saving grace. Even though Bob Adriance's family of four, including a teenage daughter, had to share a single bathroom during the remodeling of their Virginia home, they managed because, as he put it, "the children were good troopers. They were excited because they were going to get nicer rooms. It's exciting to come home at the end of the day and see what's been done."

To the greatest extent possible, get your children ebullient about the renovation. Show them the drawings and models. Let them have a hand in the design, too, by asking them what kind of room they'd like, where they'd like the bathroom, what color paint they want, and so forth. Have them accompany you during your regular inspections so that they can see the progress as it occurs and can feel that events are getting closer to completion. Let them help in any way they can.

Dust

Dust like you've never seen before. Dust like you couldn't possibly imagine. A Martian windstorm tearing through your house, pelting dust into bodies, beds, and appliances. Dust that clings with the force of a black hole. Dust that is eternal.

Get used to it. There is going to be more dust in your house or apartment than in the rest of the universe. The ordinary laws of matter and energy are boldly defied during renovation: Dust is created and no dust is destroyed.

"The dust was horrible," Bob Adriance says of his renovation. "We stayed in our bedroom. We had one room and it was livable. We retreated to that one room." The dust and debris got so bad toward the end of the project that Bob's family had to move into a friend's apartment for two days. "The apartment felt elegant in comparison," Bob recalls. "When I got back to my house it was like moving to Tijuana, Mexico."

Clean As You Go

If your entire house is being renovated, there's nothing, repeat nothing, you can do to keep it clean, at least as your mother would define the word. But if only part of your house or apartment is undergoing remodeling, then not only can you keep that part clean, but you should. Undoubtedly, you've heard someone say, "It makes no sense to clean up during the renovation because it's only going to get messier." While true, this statement is also a false philosophy. Much of the dust and debris that you ignore now will go elsewhere—and elsewhere is where you don't want it to go. Accumulated dust will venture into your heating ducts, behind bookcases in other parts of the house, between cracks in the floorboards, into electrical outlets, on the upper sides of ceiling fans, into stoves, and just about everywhere you can't get at it easily. Left alone, you may never be able to get rid of this dust. By vacuuming and sweeping dust as you go along you will limit the dust's ability to go everywhere. It's particularly important to keep dust out of vital systems such as your HVAC (heating, ventilation, and air contitioning) unit, appliances that can tolerate only so much dust before they stop working (including computers, VCRs, and ceiling fans), and especially your lungs. To give you an idea of how much dust is kicked up during a renovation, most new HVAC systems come equipped with an electrostatic air cleaner that requires the owner to clean the filters twice a year. After a renovation you must clean the filters once a month for half a year! Just keep in mind that dust likes to go where you don't like it to go.

Cleaning daily is vital if anyone in your house has allergies or breathing problems. If this is the case, make sure you have an industrial vacuum cleaner.

Keeping your house clean and neat encourages subcontractors to be a little neater themselves (relatively speaking). Stacking wood, pipes, cabinets, and other supplies has a positive psychological effect on workers, as does the daily sweep and vacuum. When these items are left in a messy state, the workers figure that it's okay to leave the rest of the house that way too. There's no chance of subcontractors being neat if you demonstrate that the house should remain a wreck.

> **Buy a couple of brooms, dustpans, mops, and pails. Leave them in obvious places. Some workers may actually pick up these tools and use them.**

Cleaning is really up to you, not the contractor. Contractors aren't good at cleaning, and it's not what they like to do. Sometimes the cleaning-subcontractors they hire aren't good at cleaning either. Eventually you will have to do one giant cleanup at the end of the renovation. Cleaning as you go along will make the task easier, less expensive, and more thorough when you do your end-of-the-job cleaning.

If your contract specifies that the contractor sweep and vacuum daily, hold him to that deal. Make it as important as completing the actual work, and tie daily cleaning to his being paid. Do not accept the contractor's plea and promise, "It won't make a difference if any vacuuming is done now, there'll just be more dust. Better if we hold off cleaning till the end." Better for him, not for you.

Cleaning up as you go along should consist of several components. The best tool you can have is a broom. Sweep that dust away! Damp sweeping is effective too. Use a vacuum cleaner to supplement your sweeping, but not just any vacuum cleaner. Household vacuums get ruined by fine plaster dust and the sheer quantity of dust these machines are asked to consume during a renovation. Obtain a commercial vacuum, which your contractor can supply. Insist that he have it on site all the time; tell him that you'll use it if he has it around.

Murphy's Oil Soap and a sponge are good weapons. They're pretty useful in helping to wipe away grease, spilled adhesives, drops of congealed plaster and other mistakes before these mistakes have had a chance to become permanent. Likewise, a sharp straightedge razor can help get rid of stuff that doesn't want to come off. Later on your contractor may get around to undoing these misfortunes, but later is when they will have hardened. Paint remover, spot removers, isopropyl alcohol, ammonia cleansers, and the old standby, turpentine, are also useful if you're interested in preventing spills from becoming permanent. A strong pair of rubber gloves makes the task easier on your hands—and safer.

Altering Your Perception of Reality

Actually, a glass of wine or two at night helps. But more than that, you are going to have to modify your conception of the universe for the duration of renovation. If you are doing serious renovation, then you need to adopt an extraterrestrial view of your home. Let me explain.

Perhaps you're renovating 25 to 40 percent of your home. For the duration you have to redefine what your home is. Instead of thinking of your house as a four-bedroom, two-and-a-half-bathroom house with living room, dining room, kitchen, and basement, you need to perceive your house as a one-bedroom apartment with a shared bathroom and a kitchen comprising a microwave, a sink, a toaster, and a refrigerator. By emotionally altering your view of your house, you will be stronger physically and psychologically. Mentally segregate yourself from the construction.

A portable, battery-operated television is a handy device to have during your renovation. It lets you have company as you're shunted from room to room.

Create artificial extensions to your house. If you once had a rec-room basement that's now become cluttered with pipes, planks, and plastic, take your family bowling once or twice a week. Imagine the bowling alley as a temporary replacement for your recreation room. Haven't been to a miniature golf course in decades? Now's a good time. There is tremendous anxiety and aggravation associated with remodeling, but there's plenty of opportunity too. If, miraculously, the workers have left your yard intact, garden! Grow tomatoes and invite yourself over to friends' houses for dinner (lunch and breakfast, too, if you know them well). Read—not at home but at the library, where it's nice and quiet. There are many ways to keep your sanity and take your mind off remodeling besides watching "This Old House."

How to Build an Addition

If you are building an addition, always build the addition first, and completely close it in before you start breaking through the wall to open it up to the rest of the house. Why? First, security for your personal possessions. Second, if you are doing the addition in winter, it's better for heat retention because you don't have a side of your house exposed to the fresh air. Third, the rest of your house will be neater. The workers will want to break through the wall as soon as possible, but that's a bad idea for you. Make it clear in the plan that the addition doesn't get tied in until the last possible moment.

Mental Deadlines

One of the worst steps you can take is to have a deadline for when the renovation must be completed, like Beth Farker did when she planned a Christmas dinner for three months after her kitchen renovation began (two months allotted for the work, as promised by the contractor, with a one-month safety margin) that was a fantasy.

Setting a deadline gives you something BIG to worry about; it makes time a crucial variable. You already have something BIG to worry about: your house. Depression and anxiety, manifesting themselves in real, physical symptoms, typically result from deadlines. Of course you want the contractor to feel that he has a deadline, but you don't want to get stressed out, as an often-heard term describes it, by relying on some good ol' boys to meet the deadline. Sometimes you can't help but have deadlines—for example, if you must be out of your old house by a particular date. In such instances you must mentally prepare yourself very, very well.

When your job involves making and keeping deadlines, home renovation will drive you, in the words of Michele Sands, a veteran home renovator, "crazy. [The contractors] just don't seem to think in deadlines." Michele's husband adds, "They'll spend days with computer programs, determining schedules. Their schedules are worthless." But, to the greatest extent possible, don't create any unnecessary time pressures. Don't

Plan a party,

Plan a wedding,

Make babies,

Order furniture to be delivered prematurely (not that it's going to arrive on time, anyway), or

Plan to get your house ready to sell for the spring market.

The Stress of Command

Hiring a general contractor is not the same thing as having a general contractor. You may find out too late that the contractor you're paying good money* to is a lazy good-for-nothing. But even when you have a contractor who's good, there will be times when you must take command. The G.C. can't be

---

*In case you've ever wondered what bad money is, here's a definition: Bad money is money you must spend before the IRS can tax it. Bad money is becoming very scarce and is worth about three times the value of good money.

there every minute; and besides, you are probably more fastidious about how windows, fixtures, and appliances are installed. Expect to plunge into the role of general contractor at the most unexpected and inconvenient times.

Being your own general contractor when there is supposed to be a real, live G.C. on the job is as stress-producing as any other aspect of renovation, in part because it's unexpected, in part because you certainly have something else to do, and in part because you're not trained for the job. "They never managed the project. We managed it," Michele Sands says about the renovation of her Michigan home. Michele and her husband, Harry, never anticipated that no one would take charge of the subcontractors. "This is why we were so exhausted. We both have demanding jobs, working forty-five or fifty hours a week, plus doing this sometimes for two hours in the morning before work, then at work during the day calling and checking, and then again in the evening. There's constant worry about who's going to be doing the job, the quality of the work they're doing, and how people are going to be paid."

"They weren't helping out with the glass order [the ceiling in the Sands' house was to be entirely glass], so it really fell to us and the architect to work out the details." Harry adds, "You'd better have a lot of time to do this and be very interested in construction."

Then There's Always Money

Money causes worry almost as fast as anything. Worrying about the financial solvency of the contracting firm, and worrying about the escalating costs of the renovation. "Almost daily there have been so many ups and downs in this project," Michele Sands says. "The financial problems of the company were one reason. When one of the subcontractors wasn't being paid, he came to us for the money." Worrying about the contractor's financial solvency is among the most stressful conditions of home renovation. Not knowing whether they will be able to complete the work, wondering every day if a mechanic's lien is going to be filed against your house is almost as bad as having dust storms swirling about. You may lose a $20,000 deposit you placed on the work, then have to find someone else to complete the project who won't give a warranty. (A crew that takes over where the work stopped won't guarantee the work because their work is founded on someone else's mistakes.)

When the costs of your project seem to be increasing, you can count on your blood pressure rising in tandem. Later on in the book I talk about what you can do to control costs, but for now let me just mention that your project will cost more than you expect. Unanticipated problems ("Where does this pipe come from?"), last minute changes ("No, better to put the door here"), adding to your wish list ("Let's put in a ceiling fan instead of a plain light fixture"), labor problems (the low-bid electrician just moved out of town)—these are the sort of things that will happen to you. You can cope with this by allocating a reserve between 10 and 30 percent of the estimated cost of your project. Don't start a renovation with exactly enough to complete it, because that amount won't be sufficient. If you have to, scale down your project to create a reserve. The extra cash will protect your psyche like an automobile airbag during a head-on collision.

Oh yes, and put enough aside for furniture.

Diary of a Home Renovator

Although there's no such creature as a typical renovation, Chris Zelkovich's experience serves as a good model of what happens during remodeling. He described his family's ordeal this way:

THE UPS AND DOWNS OF A MAJOR RENOVATION*

May 16, 1988: My wife, Liz, calls John Carley of Carley & Phillips, an architect she'd met through her job as editor of a homes magazine. He agrees to take a look at the house in June and says that if all goes well the job could be finished by Christmas.

June 12: Carley sees great possibilities. Unfortunately, he also sees great amounts of money: about $120,000—more than double what we had been considering. After regaining consciousness, we

*This diary originally appeared in the *Toronto Star,* (October 8, 1989): p. E1. Copyright © Chris Zelkovich. Reprinted with permission.

agree to negotiate when Carley says he thinks he can come up with a plan in our price range.

Aug. 7: Carley delivers the plans. Unfortunately, seeing the drawings gets us hooked on remaking the whole house. He proposes rectangular additions to the kitchen and living room, but we lean toward his optional plan, which calls for the additions to be triangular. We nickname it "The Prow."

Aug. 10: We watch *The Money Pit,* a comedy about a renovation that turns into a nightmare. Outrageous, we think. We hope.

Aug. 15: The bank tells us we can afford much more than we thought and we decide to go ahead, opting for the triangular additions. We sleep fitfully that night.

Aug. 23: Carley informs us that in Mississauga building permits take not two weeks as expected, but eight to twelve weeks. That means we spend Christmas in East Beirut.

Sept. 8: We begin the painfully slow process of searching for kitchen cupboards, floor tiles, and appliances.

Nov. 4: Carley tells us the final plans are nearing completion and will likely go to tender November 7. The building permit is either in Mississauga's zoning department or on the planet Neptune.

Nov. 24: The good news is the bids are in. The bad news is they're only $2,000 apart and about $20,000 more than we had been expecting.

Nov. 29: We choose the bid by Dave Simpson Construction because it was lower and because his Jaguar looked like it needed a paint job. By dropping a few items, we manage to save $10,000.

Dec. 12: I pick up the building permit and get a little surprise. There's a $686 charge for the permit and a $450 deposit in case we do any damage to city property. Merry Christmas to them, too.

Dec. 21: Excavation starts at 7:30 A.M. One hour later, Dave and his assistant, Jack, who are to become members of our family, hit an old septic tank that no one knew existed. Removing septic tanks is not cheap.

Jan. 6, 1989: Work starts on the framing and laying of floors. We notice the first signs of dust dunes forming in the living room.

Jan. 13: Dave and Jack begin cutting brick, stripping off the outer wall of the house. Breathing brick dust to the tune of a screaming power saw is only slightly more enjoyable than scaling a barbed-wire fence naked.

Jan. 16: This is it: the moment we feared most. Dave and Jack don't show up. We wonder if they're pulling the old contractor's trick of disappearing for a month to work on another job.

Jan. 17: We find out that Dave is in bed with a bad case of flu. We tell him how happy we are to hear that.

Jan. 27: Windows are installed as we dismantle the last vestiges of our old kitchen. We begin living full-time in the recreation room, where we'll spend much of the next four months with our microwave eating dust-coated dinners.

Feb. 14: The back wall of the living room is destroyed, leaving everything open to the elements. We wonder what life would be like if it weren't 8°C.

Feb. 27: Bad day all around: First we're told the dining room wall needs to be replaced because it's so badly rippled. Then the contractor tells us the ceramic floor tile is delayed indefinitely because of a strike in Italy.

Feb. 28: More bad news. The wiring system in the house is only 60 amps and we'll need 100 amps to accommodate the air conditioning and new lighting. Cost: $800.

March 2: Appeals to the Italian government fail, so we look for new tiles. We realize the white suit in my closet is in fact my blue suit.

March 6: After hours of searching, we choose a new kitchen tile. We discover I am not going gray, but am just covered in plaster dust.

March 9: Contractor persuades me to tear down the old living room ceiling myself—thus saving about $400—and have a new one installed. Unaccustomed to saving money, I agree.

March 10: I note that kitchen cabinets are two days late. Kitchen company gives numerous excuses, though it doesn't mention any Italian strikes.

March 12: After spending the weekend tearing down the ceiling, a task highlighted by a shower of loose-fill insulation, I decide I would have gladly given someone $400 and my car to do this.

March 16: Disaster: Removing the living-room ceiling reveals that the upstairs shower has been leaking slowly behind the tiles for some time. That leads to the discovery that the wall beside the bathtub is saturated. Tiles have started falling off. What next?

March 18: The kitchen cabinets arrive and are stacked up in the kitchen. We tried to dust the furniture today, but the wheelbarrow couldn't handle the load.

March 20: Kitchen installer tells us he'll be finished in two days.

March 22: Installer returns and announces that the job may take a little longer than anticipated. Only later will we realize what a funny man he is.

March 27: Siding is done on living room addition. Meanwhile, kitchen installer enters his second week on a two-day job.

March 28: Steps are installed outside French doors. Dust storm generates in bedroom.

April 3: As the kitchen comes slowly together, we notice that several doors don't fit and that the countertop is bubbling up. We realize this may not be the way we ordered it.

April 5: The architect inspects the kitchen and presents a list of woes to the kitchen company. To our surprise, they agree to replace the countertop and several other items that were obviously made for a kitchen in Australia. It will take two weeks, they say.

April 17: A major advance: the kitchen lights are connected. It's beginning to look like a real kitchen, except for the missing countertop, doors, gables, door pulls . . .

April 21: Kitchen people promise all will be completed Monday. Dare we believe them?

April 24: Miracle of miracles. Kitchen installer arrives first thing in the morning. But we return home to find that he installed about a dozen door pulls and then left.

April 25: We return home to find that the installer hasn't shown up at all. We discuss plans to firebomb the kitchen showroom. We find the missing coffee table under mound of dust.

April 26: Contractors return to do some painting. Meanwhile, kitchen people promise to install everything by the weekend. We laugh hysterically.

April 29: A busy day. Kitchen installers deliver the new countertop. But it takes them all day to install it and we're still missing four doors, two sets of shelves, a kickplate on the island, and a coatrack.

April 30: For the first time in months, our house feels like a home again. To celebrate we break in the kitchen by chopping vegetables and making a fruit salad in our kitchen.

May 6: We notice a strange atmosphere in the house, something almost frightening in its scope. Realize later it's the absence of dust.

May 19: Architect and contractor conduct a "final" inspection to verify that 97 percent of the work is complete. A few minor details are unfinished, but other than that it's all done.

May 20: When we arrive home, we find a note explaining that the kitchen installer didn't have the right doors to finish the job. He says he'll be back in four days. We do research on making Molotov cocktails.

May 23: Much to our disbelief, the kitchen installer shows up and finishes the job.

May 26: We hold a celebration party for the contractors and architects and their wives. The party is held five months after construction started and more than a year after we began the project. But the consensus is it was worth every dollar and every minute. We basically have a new house.

July 7: We take a dust bath, for old times' sake.

The Kitchen and Other Forbidden Places

Do the workers have access to your wine collection? Can they rummage through your refrigerator? Consume your son's school lunch? Ask your daughter out on a date? You should make it clear if they can't. Also, can they store their Cokes in your refrigerator? Eat at your kitchen table (while reading your newspaper)? If your kitchen is not undergoing renovation, it may be the only sanctuary in the house: It's certainly a place you don't want invaded by workers who leave lingering aromas and palm prints. Most workers shy away from using the homeowner's kitchen other than to get a glass of water or keep a bag lunch cool, but it's best to set the ground rules for using your kitchen at the onset. Do this through the general contractor; he should communicate to his subcontractors these ground rules.

If no family member is going to be in the house in the daytime during your renovation, it's a good idea to disconnect your phone number from long-distance services. You'd be surprised how many "suppliers" tradespeople have to call in various parts of the country. Isn't the Dominican Republic a good place to look for tiles? You can always make your own long-distance calls through an operator and bill the calls to your credit card number.

Which bathroom should the workers use? Designate one for them, and make it clear that they should not use any other. Keep that one bathroom stocked with all the necessary supplies such as soap (but don't be surprised if the soap bar doesn't shrink). Paper towels are better than cloth ones for that bathroom, because I doubt you'll want to ever use the workers' towel again.

The Might-As-Well List

As you're suffering through dust, noise, loss of privacy, and wholesale assault on your checkbook, what difference does it make to do just a little more work on your house?

Yes, indeed. What difference at all? The disruption of your life can't be any worse.

In truth, every bit more that you renovate takes more time, costs more money, and disrupts your life more, but this incrementally greater disruption is nothing compared to doing an extra, independent project later. My philosophy is that you should do as much now as you can fiscally afford. You may never want to travel through remodeling land again.

The extra mess that a little more work is going to create pales in comparison to starting anew later. Besides, the workers are here—now; the apparatus to protect your floors is here—now; you are used to the mess of renovation—now. Later will be a different story.

At the end of the renovation you won't want to go through it again, and, consequently, you may never get that beautiful pantry you're thinking about.

Just because you have plans doesn't mean you can't change them. In many cases, you can add to what has already begun, always at a price—but practically anything is possible during remodeling.

What should you put on your might-as-well list?

Whatever you want to do—both big and small projects, especially the small ones, because later you may decide it's not worth it. The list can include projects that help improve your house's energy efficiency or insectproofness. Walk around the house and decide what could be made better. Try and anticipate your needs a couple of years from now. If your might-as-well list is short and fairly simple, you can coordinate it as a second, independent renovation that happens to be going on simultaneously—for this project you are the general contractor. The G.C. who's overseeing the main work doesn't have to get involved in this co-renovation, as long as the two don't conflict.

To inspire you, here's a list of the kind of work that you can initiate toward the end of your principal renovation:

Add more electrical outlets.

Insulate walls; sound-proof walls.

Install a security and/or a hard-wired smoke detection system.

Install vents or an automatic fan in the attic.

Replace a stall shower with a bath/shower.

Fireproof the wood beams around the furnace and hot-water heater.

Put in a new main water pipe to your house to improve water pressure.

Replace an old toilet.

Replace broken storm windows and screens.

Add an air-cleaner to your central heating system.

Put a through-the-wall air conditioner in an attic room.

Install a ceiling fan.

Build a cedar closet in the basement.

Parking

Who owns your driveway? Most of the time it's you (except for those nights when half-blind louts can't tell the difference between the street and your driveway.) However, when renovation is in progress, your driveway (and the surrounding area) will look like a used-van lot. Workmen will park where they want to, kind of like the old joke: "Where does a lion sleep? Wherever it wants." There's not a whole lot you can do to keep the workers' trucks out of your driveway, especially if that's the only place they can park. You should indicate your preference, however, that they not park their vehicles on your lawn or in your neighbor's driveway.

Don't Blow Up Your House

It happens. Really does. Paul Locher, a builder, was working in Pittsburgh and got an emergency call in the middle of the night that a furnace was exploding. Rather, was making exploding noises—serious and potentially dangerous. "So at 3:00 A.M. I went to this woman's house. I went into the basement, and took the face cover off the furnace to examine the unit." At the same time, the woman went upstairs and turned up the thermostat. "Naturally," says Paul, "there was a large explosion not contained by the faceplate. A flame shot out ten feet and the force picked up my two hundred pounds and threw me back fifteen feet. The lady came running downstairs yelling 'Did you see it, did you see it?!' She wanted me to see the problem." Paul did. (For technically curious readers, the alignment of the burners wasn't level, so gas was filling up one side of the combustion chamber faster than the other. When the furnace lit, it lit off balance, correcting itself by burning all the fuel at once.)

If you want your contractor to live through the renovation—and if you want your house to survive—don't play with switches, outlets, fuse boxes, thermostats, HVAC controls, faucets, or anything else while work is in progress without asking first. That funny container in the kitchen may be a propane tank—this is not the time to adjust your new gas range. Don't test appliances without checking that it's safe to do so. If you turn on an air conditioner you may blow the circuit that the contractor is using for his saw, and if the saw stops in the middle of cutting wood, the wood could bind and there the saw will stay. Running water into your new shower might seem like a good idea, but it isn't if the plumber hasn't tested the connections on the drain pipe. Don't willy-nilly reconnect the main power to your house if it has been turned off. Consider your house a dangerous place, a foreign place, while work is going on. You are an alien in a strange, new world. Presumably you will have your areas—the places where no work is being done—where it is safe to use electricity.

The same thing goes for areas where adhesives have recently been used. Tile floors need time to harden. Paint needs time to dry. Plaster needs time to set. Floor varnish needs time to dry. Throughout the renovation the areas that you, your family, your pets, county inspectors, and *other workers* should avoid are going to change. Have these areas clearly labeled, and barricaded. For tilers, it's second nature that nobody should walk on a tile floor until it's ready, so the tiler may not bother to tell you. City electrical inspectors don't care one bit about tile floors and will waltz right over them to get to their destination, an electrical outlet. This happened to one New York homeowner: To verify that one outlet was working correctly, every fourth tile now looks like it has endured a California earthquake.

You should ask where it's safe and unsafe for walking. You should ask what's safe to use, and what's not safe to use yet.

Resist the temptation to start enjoying your new house before the suffering ends.

Be Angry

Rage has a place in every renovation. The swells of anger can rapidly reach tidal proportions. Knowing when to release your anger and how to direct it is an important renovation survival technique.

In Chapter 9, "The People Who Do the Work," I list some of the excuses typically heard from contractors—excuses they offer as reasons why someone can't work today. "My truck broke down" and "There was an emergency wedding" are common. Sometimes the consequences of these excuses are benign if an-

noying—nobody shows up to install your kitchen cabinet. But sometimes these excuses lead to major problems, as when a Calgary, Canada, homeowner was left with a canyon-size hole in a wall for several weeks because the subcontractors kept giving "reasons" why they couldn't show up. Another Canadian homeowner kept getting water in her basement through an improperly installed window, and only excuses from her builder.

These are the kinds of instances when you should not tolerate excuses. If you've tempered your aggression, unleash it now. When a contractor keeps giving you excuses and the problem you have is major, take off the gloves and let that contractor know that you want *someone* to fix the problem NOW.

If you have to get mad, get mad at a contractor, not your spouse, friends, children, or pets. Direct your anger appropriately.

2 | Should You Use an Architect?

When to Use an Architect

You've decided to renovate your home or apartment. And so the question that naturally ensues is, Why?

Perhaps it's because you *must* spend $75,000 right away and you already own a Mercedes. Perhaps it's because you feel you've led too pampered a life and should suffer. It could be that you have lots of checks that you want to part with because you're changing your checks' design. Or maybe it's because renovation is the only thing that will drown out the sound of your teenager's drum playing.

While these are all rational reasons for remodeling, eventually you will have to step beyond the jubilation of anticipating great noise, dust, and expense and actually decide what to do where. When you reach that stage, you'll probably wonder, "Where can I find a good architect?"

This is not the first question you should be asking. The first question, really, is, "How should I plan my renovation?" One very good way to think about your renovation is to think about whether to use an architect. Deciding whether to use an architect will tell you a lot about how to structure your plans.

Not all remodeling projects require an architect's skills. In fact, many renovations are better off without an architect's assistance.

26

Architects are expensive, don't always understand the particular engineering and construction matrix of your home, aren't always well equipped to translate your needs into a plan, and won't always be on site during construction. What architects are often good at is attempting to coax you into accepting the kind of design that they were taught in architecture school or that they like most.

What criteria should you use to decide whether to employ an architect? That's not an easy question to answer. You can always have an architect visit your house; there's certainly no damage done, except sometimes some money spent. But money is a scarce resource in most renovations, not to be squandered getting frivolous opinions. I can't give you a precise formula to determine whether you need an architect. Before offering my general rules of thumb, however, let me suggest one reason not to use an architect. Don't think you have to hire an architect to draw a plan. A competent builder can draw a plan. *You* can draw a rudimentary (or sophisticated) plan, as you'll see a little later.

By thinking about the kind of space you want, you should have a sense of whether you need an architect's services. If you know what you want where, you may not need an architect to validate your wishes.

Generally, small jobs—replacing fixtures in a bathroom or kitchen, closing in a screened porch, finishing a basement, installing a through-the-wall air conditioner—don't require an architect's services. Even building an addition that uses off-the-shelf plans, or a builder's plans, doesn't require an architect's skills. Beyond these kinds of work, whether to employ an architect becomes complicated.

At the onset of your renovation you should decide what kind of remodeling you want. This decision goes hand in hand with determining whether you ought to hire an architect. There are four basic kinds of renovations:

Category I.

Air—not designed—space. A renovation that just creates extra space—a room, deck, or usable basement with no fancy frills.

Category II.

A renovation that's concerned with inside and outside appearance, but that isn't going to appear in *Architectural Digest.* Something *you* like to live in and look at. Practical aesthetics, yes; glamour, no.

Category III.

A renovation that can be published. A renovation that architects would call "architecture," not merely a room.

Category IV.

Any of the above that involves remodeling oddly shaped, cramped, or otherwise geometrically complicated spaces.

If you are a Category I renovator, you probably don't need an architect. Category I involves creating space, nothing fancy; just more space for the purpose of having more room. Examples include a basic deck or adding a bathroom. People who are doing Category I renovations are primarily interested in more living space.

Category II renovations are a little more sophisticated. If you consider yourself interested in design, in the way things look, then you may want an architect to help. Category II renovations tickle your intellect as well as give you extra space for a laundry machine.

Category III renovations are the stuff you see in the Sunday glossy magazines that newspapers publish. For these renovations—architecture for architecture's sake—you need an architect.

Finally, there are Category IV renovations: Any of the above types of renovation that involves remodeling oddly shaped, cramped, or otherwise geometrically complicated spaces, or involves zoning or historic district considerations. Category IV houses often require an architect's unique vision to manipulate the space into something you can use.

If you like formulas, this is as good as any.

Washington, D.C., architect Ted Fleming, proprietor of the firm Architectnique, likes a more intuitive approach to whether you need an architect. "I don't think it's too complex where you need an architect or where you don't," he says. "You can get projects that are incredibly well built but not well planned. Traditionally, you look to an architect to look out for you.

"With a lot of work that's done on a small scale, you don't really need an architect. If it's minor enough, to involve an architect would be top-heavy, too expensive. For example, to remove kitchen cabinets or replace a window—a talented builder can handle that. But when the homeowner is in the position of having to have several tradesmen in and manage them, that's when you need someone to plan.

"When it comes to any issue of planning or *design* you need an architect," Ted concludes. "If you are working within an existing space and simply upgrading, you don't need an architect. If you want a design, you don't get that with a builder, even with a talented builder."

Speaking critically (I have nice things to say about architects later), architects generally have four goals in mind when it comes to home renovation. In order of importance, they are

1. Win an award,
2. Have the project photographed in an architectural magazine,
3. Make money, and
4. Get a good recommendation.

There's certainly nothing wrong with any of these goals, but keep in mind that winning an award, while helpful for the architect, may not give you the living space that you want. Take a look at the homes and apartments featured in *Architectural Digest.* Would you want to live in one? Using an *Architectural Digest* plan would, more than likely, mean that your life-style would have to conform to the space, rather than the other way around.

The actual decision of whether to use an architect can become complicated, and skirt these neat formulas I've presented. For example, another reason you might want an architect is when you want to take into consideration the neighborhood, the house's site, the existing style of the building, and the history of the building. In a historic house, an architect may be a must if you want to preserve the house's historic elements while still gaining space. Designated historic districts often have specific architectural requirements that must be met by home renovators.

Ultimately, if money isn't too much of a consideration, it never hurts to engage an architect's services. You will always get something out of the plan—probably something good—and there's no law that says you have to use all, or even part, of the architect's plans (or consultation). Plans are paper, and paper can be erased or burned easily. (Beware, however: Some architects' contracts give them the right to retain the plan if you decide not to use the architect's services.)

Architects Aren't Always Swift

Architects like to have the entire renovation planned in advance and like to take a long time—up to a year or more with some architects—to do the planning. If this is your preferred style, then an architect will suit you better than a builder who is working on a more flexible plan-a-little-and-decide-as-you-go schedule. "Ideally, everything should be decided and put on the drawings before the project," says Washington, D.C., architect Robin Roberts. "It's worth the wait. You can only be disappointed if you decide as you go along." Although, intellectually, having everything determined in the plan sounds right, it isn't always best. Chapter 7, "Planning Ahead and Keeping to Your Plan" goes into more detail about whether you should plan everything ahead of time or make decisions as you go along.

Another way to evaluate architects is to talk with builders or design-build firms. Like architects, builders (or contractors) and design-build firms will give you an impression of what they would do with your house while they walk through it. Have the builder sketch a plan. Are you content with the builder? Do you like his ideas, or do these ideas seem limited? How much you like a particular builder's ideas and plans will help you determine whether you should use an architect.

If you move ahead with the architect, compare the builder's and the architect's plans. Do you see a significant difference? Which do you like more? Which will cost more to accomplish?

As have millions of homeowners, let's say you've decided to use an architect. How do you find one? Word of mouth is the best method. Ask friends, colleagues, and acquaintances how they

liked their architect. Did she do a good design? Equally important, could the builder implement this design? Was the design
flexible enough to cope with unforeseen structural problems?
Did the architect oversee the project? When you have the names
of some architects, examine their work. When you view a house
or apartment, be sure to ask whether the architect designed the
part you like (as opposed to the homeowner coming up with the
idea). House tours held in neighborhoods across the country are
superb ways to see examples of architects' work (as well as to get
ideas for free; after all, why else would a particular house be
selected for a tour if it didn't have a complex or compelling
design?). Most neighborhoods hold their tours in the spring or
autumn; take as many as you can find.

You might get some help from the local American Institute of
Architects chapter. The AIA will recommend architects to you.
Before you use any architect, even one recommended by the AIA,
inspect the architect's work and interview former clients.

Once you've narrowed the choice of architects to a few, interview them. In your house. Describe your life-style, budget, time
frame, and wish list and ask, What would you do? Pose a problem
that you have with your current space and ask the architect if he
sees a solution. Ask, Are my expectations realistic? Ask, Do I
need an architect? Inquire about how well the architect keeps to
budget, how well he keeps to time frame. Ask the architect how
he feels about overseeing work, or whether he prefers to delegate
that job. What sort of physical tradework (carpentry, electrical,
etc.) has the architect done himself, if any? Ask the architect what
kind of project he likes best. Spend an hour or two getting to
know the potential candidates. It's important that you and the
architect get along and that you trust the architect to translate
your living needs into a plan.

Pose each of these questions to homeowners the architect has
worked for. Uncover their opinions. References are vital when it
comes to selecting an architect. Don't remodel your home without them.

Coordinate Your Architect's Plans with the Builder

The relationship between your architect and your builder is vastly more important than your relationship with your in-laws. It's crucial that the plans work, otherwise they're not of much practical use—and potentially disastrous. (Plans that don't work because of unforeseen problems can require that certain work be undone and started over again.) In fact, before you pay the architect his final bill, it's worthwhile to take the plans to a builder (you should already have been coordinating with one by this stage) to determine, first, if the plans are technically feasible and, second, whether the plans are within budget. Even if everything in the plan is feasible, the execution may cost many times more than you expected.

Realistic plans are the only plans that are of any use to you.

Plans without Architects

You, the homeowner, are without question the best person to conceptualize your own design plan. I'm not talking about actual blueprints, which you may eventually need, but a rough idea (or sketch) of how you'd like your house transformed. Nobody knows what you want your house to be like better than you do. Nobody knows your present life-style and future needs better than you. Every architect and every builder comes to you with preconceived notions of what looks best to them, and unless you enunciate your wishes, *their* notions will become a part of *your* plan. So, get out a pencil and start jotting down ideas—and visual aids, if you have that talent.

HOW TO DRAW YOUR OWN PLANS

Drawing your own plan has nothing to do with drafting in scale inches. It has nothing to do with drawing in little couches, tables, and beds. It has nothing to do with figuring out how to mark where outlets and switches should go. Neither does drawing your own plan have anything to do with being able to recreate an existing floor scheme on paper.

The first step in drawing your plan is to walk through the space

you want to renovate, pretending that there are no walls, no ducts, no fixtures, no doors, no windows standing in your way. Pretend that the space is a complete blank. (If you know where the load-bearing walls are, take them into consideration*; otherwise worry about load-bearing walls and other crucial structures later.) What would you put where? Would you move walls around, if you could? How about windows and doors? Pretend the inside of your house is a tabula rasa. What shape and size would you make your bedroom? Would you use part of the bathroom for a walk-in closet? How about moving the basement entrance to create a pantry? Anything is possible in your imagination! Can you visualize putting the bathroom sink where the toilet is? In your mind's eye, move the house's front door to create space for a front hall closet. Stimulate your imagination. Not only is imagination a free tool, but it's the most powerful tool you have.

Next, write what you want in words. If you're not sure what you want to do (or if what you want to do is feasible) then write out several plans and label each of them (Plan 1, Plan 2, etc.). There's even an advantage to writing your plan rather than drawing it: For nondraftsmen, written lists are sometimes easier, faster, and more accurate. In addition, because drawn plans have a language of their own, unless you draw your plan in that language there's a risk that the contractor will misinterpret your plan. Drawing your own plan if you don't know the nuances of the language is like trying to survive in France on a semester of high school French: You'll manage, but you may make some unforgivable mistakes along the way. Naturally, your ideas should be clear and understandable to the contractor. Because you're not familiar with the language of contractors, keep your written thoughts simple, just as if you were talking in a language you didn't speak well.

It's fine to repeat information in your written plan. Your writ-

---

*Hint: Joists run opposite floorboards, so when you are facing along the length of the floorboards, the load-bearing walls will be to your left and right. This formula works only in open spaces with main walls, not in closets, oddly shaped structures, and other difficult areas.

ten plan should include notes on how you expect to use the redesigned spaces. The more your architect or contractor knows about your style of living, the more he'll be able to tailor the construction to suit your needs. This is another advantage of a written plan to a drawn plan: You can explain why you want something done or mention what's going to go in the space after the renovation. You can—and I hate this word—*prioritize.* Think of this list as your wish list.

As you write your plan, consider how you live. How does the traffic flow though your house? What are your future plans for children, visitors, retirement? (For example, a third-floor bedroom suite can sound wonderful when you're in your forties, but consider that you might be living in the house when you're in your seventies.) Don't plan an ideal house, plan an ideal house for you. Unless you anticipate moving within the next few years, don't plan the renovation as a way to increase the value of your house, plan it for yourself. Be analytical, critical, whimsical, logical, imaginative, and innovative. In a broad sense, consider your budget, too, but don't worry about it at this stage. Some things you can guess are going to be too expensive, but let your builder give you limitations; don't impose your own. What is too expensive can be modified when the real plans are drawn.

Designing a renovation has much in common with other sophisticated forms of problem-solving. It's both a highly logical and highly intuitive process. Focus on your problems with the current space—that's the best starting perspective. Think about the global aspects of your renovation (where to move walls, how large to build a kid's room), and the details (what kind of sinks, what shape tiles) will follow.

From the moment you decide you want to renovate, you should be clipping pictures from magazines and visiting as many other houses and apartments as possible. You just might run across a picture for which you are able to say, "We want that bathroom."

Educate yourself about your house. Try, without straining, to determine which walls are load-bearing, where ducts run, where plumbing goes, and where the sewer line is located. How is your electricity? Is it adequate for the central air conditioning system you want? You don't have to have the complete or correct answers to these questions, but the more you know about the house the better prepared you will be for your meeting with the contractor. Often it's impossible to know with finality the structural features of a house until the walls are smashed apart by the demolishers. Ask people who might know about your house for structural information: real estate agents, previous owners, the chairperson of the local historical society.

Now modify your written plans—if necessary—in light of any structural information you have.

Here's a typical word plan:

First Floor

General

Plaster where needed.

Sand and polish all floors except living room, which should just be polished.

Paint all walls (colors to be determined later).

Add shoe molding where needed.

Build front hall closet (the only space that seems to be available is under the staircase—can one be built?).

Living Room

Polish floors.

Paint (colors to be decided later).

Seal off with plastic during construction—keep workers out!

Add phone jack.

Dining Room

Replace ceiling fixture with new fixture.

Seal off during construction—keep workers out!

Don't polish or sand floor!

Kitchen

Completely redo, replacing all appliances.

Appliances to be chosen.

Put down lights in ceiling.

Construct pantry/closet where refrigerator is now.

Put refrigerator in left corner as you enter kitchen.

Remove door (door no. 1) between kitchen and pantry, keep open.

Put sink under window (window no. 1) next to refrigerator, leaving space between fridge and sink. We like to look out the window while washing dishes.

Shorten window (window no. 2) facing yard so that counter can run evenly under window.

Build pantry/cabinet between door to deck (door no. 2) and door to dining room (door no. 3).

Put stove to the right of window and run exhaust outside.

Replace door to dining room (door no. 3) with swinging French doors.

Extend wall to right as you enter the kitchen, so that a kitchen table can be placed against it.

Add HVAC duct(s), but not against the large main wall; that's were we intend a painting to go.

Put in trash compactor.

Build cabinets to ceiling with lazy Susan at corner cabinet. Put sliding shelves in cabinets below counter.

Put electrical outlets around room.

Put telephone jack for wall-mount phone on wall to the right of door no. 1.

Run counter all the way from the refrigerator to the door leading to the deck (door no. 2).

Countertop to be tiled, floor to be tiled (will select tiles later).

Second Floor

General

Move wall between back bedroom and front bedroom three feet (or so) toward front bedroom (street side) to make the back bedroom larger. (Back bedroom will be the master bedroom; it's off the street and quieter.)

Put down lights in hall ceiling.

Redo floors.

Paint.

Shorten hall as necessary to provide space for larger bathroom.

Retain entrance to master bedroom as is.

Front Room (Closest to Newark Street)

Move wall from back bedroom (alley side) forward several feet (this is the wall that was moved to make the master bedroom larger).

This room is to be Anne's home office.

Change old door opening to compensate for the wall shift, as load-bearing wall permits. New opening may now have to be at an angle. Closet will now become part of hall, and the other little front room will have access to it. Some refiguring may be necessary.

Put in ceiling light fixtures, to be picked.

Install two telephone jacks.

Add shoe molding around new wall to match existing molding.

Fix existing wall sconces.

Master Bedroom (Room Adjacent to Sun Room)

Move wall toward front room. (King-size bed will go against this wall at the center of the wall.)

Insulate wall between master bedroom and Anne's office so that sound doesn't carry.

Install heating duct at wall opposite bed's wall. Put duct up high because a bureau will go against that wall.

Install electrical outlets and phone jacks on either side of where the king-size bed will go (bed is seventy-two inches wide with nightstands).

Put wall sconces at either side of bed (Sconces to be picked later. How high above bed?)

Build door through existing closet to next room, which will become the walk-in closet.

Install ceiling light fixture (already bought!).

Refinish floors.

Paint.

Sun Room

Replace screen with glass.

Insulate.

Even floor, and build new floor if necessary.

Open door to adjacent room, the walk-in closet.

Walk-in Closet

Cut back room in half by building wall. Side closest to sun room will be a walk-in closet; north side will become part of the bathroom.

Open room. Will get closet company to design and build later ourselves.

Put door between closet and bathroom, near where the tub will go.

(There will be three doors in the walk-in closet; one to the bedroom, one to the sunroom, one to the bathroom.)

Add ceiling fixture.

Bathroom

Take down wall between the bathroom and the north half of what is now the other back bedroom.

Jacuzzi and shower will go in east half of bathroom, Jacuzzi at corner of room (corner tub). Jacuzzi to be selected.

Put some kind of door between the toilet part and shower part of the bathroom.

Put double sink where tub is now (west wall of bathroom).

Large mirror at sink.

Move window as needed.

Push bathroom into the hall, and shorten hall behind where sink is now located. Toilet will go in this alcove (is this possible?).

Put in recessed ceiling lights.

Tile around Jacuzzi (tiles to be bought later).

Tile in shower.

Tile around mirror.

Tile floor.

Third Floor

Paint.

Refinish floors.

Put in ceiling fixture in front room (southwest room).

Other

Install metal grates around basement windows.

Replace broken shingles on roof.

Clean off asphalt roof outside attic window.

Put in new HVAC system with electrostatic air filters.

Are any more ducts needed? Any more returns needed? Note: In the summer we like it very cool, so when there's a question, we'd prefer putting in a duct.

Identify, then remove or seal, all asbestos—whichever is safer.

A design list or wish list can be as detailed as you want or need. It isn't necessarily in the order that work should be done, nor is it comprehensive. The list gives your architect a place to start. If you know what kind of fixtures you want—a Kohler Model X sink, an Amana refrigerator with the freezer on bottom, a Casablanca Satellite fan—or if you know any other materials you want—the baby's room wallpapered with brand-name animal paper or olive Crete tiles in the kitchen—say so. If you're certain that a bed is going to go against a particular wall, indicate this so that the designer doesn't put a door there. A one-page wish list is fine, but a twenty-page written design is even better. The more information you include in your wish list, the more accurate the blue-prints will be and the better the outcome of your renovation.

FROM BLUE INK TO BLUEPRINTS

The next step after writing your wish list is to draw a plan. Drawing your own plans not only can save you a great deal of money, it will give you great insight into what kinds of designs are possible. It will enable you to say to the builder or architect, "Sure you can do that. Just move this wall two feet and put the air duct here"—and show him! By drawing your own plans, you will be able to anticipate all the myriad ways you can renovate your home.

Now hold on a second—don't go for the Valium. If you can't draw, you don't have to draw. Drawing a plan isn't a requirement for either your renovation or using this book.

If you can't draw and would like to produce a rudimentary or sophisticated plan for your builder or architect, there are several aids available. A variety of computer programs enable homeown-ers to develop detailed plans that builders can readily use. You'll find that using these programs makes thinking about your future house fun. Playing around with drawings will also give you ideas that haven't already voluntarily come into your head.

If you're not computer-inclined, several mechanical drawing systems—complete with miniature couches, pianos, tables, beds, refrigerators, and so forth—are available. My wife, Peggy, used the Home Designer system available in most hardware stores. The system costs about twenty dollars and comes complete with all the instructions you'll need to create a working plan of your house. The hardest part of using Home Designer is making the initial measurements. But once that's done, you can move walls, staircases, entire rooms, at will. If you can get your architect or builder to tell you which are the load-bearing walls, you'll even know which walls can't be moved. Play with the plan. Consider all options, even those you aren't crazy about, because you don't know where oddball ideas will lead. When you've got a plan (or several) that you like, photocopy it and present the plan to your architect or builder and say, "Here, can you do this?" or, "I want to create a walk-through closet like this between the master bed-room and bathroom, but I don't see how to get around the chimney" (even chimneys can be moved, for a price). My wife and I wanted to make a pathway between our bedroom and what would become the walk-in closet. One architect suggested moving the chimney (a $25,000 idea); our builder, Paul Locher, built a door between the fireplace and the load-bearing wall for about $1500.

Home Designer isn't the only route you should take when it comes to designing your space. Visit houses (or apartments). When planning your design, you cannot see too many other homes.

Also buy those architectural and home remodeling magazines. If you spend forty dollars on magazines and find a single picture that has the bathroom of your dreams in it, that would be forty dollars well spent.

If You Use an Architect

As I've mentioned before, architects don't always have an intimate understanding of the structural parameters of your house or apartment. Oftentimes, they may also lack a realistic knowledge of how much turning ideas into reality costs. Many architects don't keep track of the current costs of pipes, new ducts, carpenter's fees, and so forth.

There is something else you should know about architects: Architects don't make mistakes. Unfortunately, this is not a good thing, as Jim Petrick and Barbara Fredericks discovered with their Maryland house. Their architect, who masterfully integrated a heating and air conditioning system into their house, put an HVAC vent in the only place that their living room couch could possibly be situated. When they pointed this out to the architect, he eventually recognized that it wouldn't make much sense to have an air duct in the living room if you couldn't have a couch in that room, so he drew up a second set of plans which created a window sill-like structure above the couch and put the vent into that space. Then he presented the couple with a $500 bill for the HVAC design.

That's how architects don't make mistakes: They revise their plans to compensate for the "adjustment" that has to be made and then insist that you pay for it.

Having said some uncomplimentary words about architects, let me save face by adding that architects can be a valuable, if not essential, ingredient in your renovation. It all depends on how much insight you have into your own plans. An architect can design your space. An architect can interpret your living style and translate that style into redesigned space. An architect can provide innovative, even breathtaking ideas about remodeling your home. Often, people who renovate their homes are too conservative—or, more aptly, too chicken—to create pioneering spaces.

Many people who hire architects also want their architect to oversee the project. Generally, this is the aspect of home renovation that architects like least. And so they write into their contracts a limited number of hours that they will be on site, and they make those hours expensive.

Most of the time, once the architect has completed the plans, he has completed the bulk of his work. This varies from architect to architect and project to project, but keep in mind that architects are not general contractors. They do not oversee the day-to-day construction. When an unforeseen circumstance dictates an on-site change—your stair landing is in the middle of a window, for example—you have to decide who plans the adjustments. The architect? You? The builder? The on-site carpenter? The builder is going to be on the job, but the architect designed the plans. This isn't an easy decision, but it's an important one.

One rule that is crucial beyond anything else when selecting an architect is: Don't give the architect on-the-job training. If this is the first time an architect has done a particular type of design, let the architect practice on somebody who hasn't read this book. An architect who's been on her own or who's just starting out professionally isn't a bad choice, especially if that architect has had experience doing what you're planning. A medical analogy: If you suffer recurrent heartburn you can become very well educated as a layperson in digestive system disorders. However, there's a world of difference between reading exhaustively about a problem you've experienced and being a practicing physician who has seen hundreds of patients with similar complaints. The experience of doing something, seeing something over and over again, gives you insight that cannot be gained from merely reading about or occasionally suffering from a problem.

An architect with little on-his-own experience but with experience doing your kind of renovation can save you considerable money. When you hire an architect, you are paying for his time, his ego, and his experience. Make sure you get the first and third items.

As you'll find me repeating throughout this book, recommendations and references are important. If you can't find anyone who says nice things about the architect you're thinking of hiring, don't hire him. It's risky to hire any architect without getting some recommendation. Listen to what people have to say about the architect. Probe the architect's previous clients; don't be satisfied with surface recommendations like "He's great!" or "His design was terrific."

Has the architect done work similar to what you need? For example, if you're renovating an apartment and the architect has only worked on commercial projects, your apartment might end up looking like an office. If the architect has only worked on new town houses and you live in a historic district of older, detached houses, the architect might not be sensitive to the historic aspects of your house.

Some architects, says Ted Fleming, "understand that the best way to gain a reputation is to build something—this is what architects did hundreds of years ago." Almost nothing beats an architect who likes to get physically involved in construction. One complaint often leveled at architects is that the profession doesn't often take into account the engineering reality of a project. Once a builder gets hold of a plan, he may have to rework it; constructing some of what the architect has drawn turns out to be impossible. If you've found an architect with dirt under his fingernails, hire him.

THERE ARE BAD ARCHITECTS—AND BAD CLIENTS TOO

In hurried flashes of deep honesty, we have to admit that as home or apartment owners engaged in the throes of renovation, we haven't been angels. We've badgered, yelled, sworn, lied, and committed other sins during the process. But which of these sins is the worst? None of these acts has made the lives of our architects easier.

According to architect Robin Roberts, "A bad client is one who views the architect as a drafting service. Drafting you can have done cheaper by someone other than an architect."

The second sin from an architect's point of view is the home-owner who expects "too much for too few dollars," according to Robin. Clients who have unrealistic expectations about what they can achieve on their budgets are a royal pain to architects. The client wants to transform a one-dollar bill into a two-dollar bill. Every architect wants to see bigger, more expensive projects too (especially if his fee is based on a percentage of the project's cost).

Some client attitudes are merely bad; others contribute to a bad project. Says Ted Fleming, "In the best of all possible worlds, there really isn't a bad client. A lack of communication skills on the part of architect and client leads to the perception that there is a bad client. What we architects do as a service isn't necessarily complicated, but it is complex. Without sufficient education of the client, people get lost at the beginning, and this builds distrust and resentment. I learned early that if you don't take time up front to explain how the process works, you're setting yourself up for problems."

Ted points out that clients often go into renovation with misconceptions about what an architect is supposed to do. "I've met people who have registered surprise that we would have to spend so much time talking about details, how a bedroom or kitchen would work. They think we have a magic wand, pull on a vast store of information, and make the thing that looks so great in the magazines," he explains. Architects—good architects—have to learn what you want; they have to explore both your current and your future life-style. If all you want is a p-l-a-n, you are going to be disappointed.

A lack of humor is a lousy thing from an architect's perspective. If you aren't willing to laugh a little, then you're probably going to take your frustrations and anxiety out on the architect. Maybe he deserved it. But not all of the time.

Architects like to see a plan drawn in the beginning and ad-hered to. They like what's on their blueprint to turn into what's in your house. Architects believe that homeowners' requests for changes (other than minor changes like tile and paint color) could be avoided if all desires were addressed in the design phase. From an architect's perspective, it's worth taking a l o n g, thorough time during the planning phase, so that the plans are perfect. Roughly speaking, a medium-size kitchen, dining room, and bath renovation should take three to five months to plan, according to one architect. The longer something takes to design, of course, the more you can be charged.

But then, architects *like* drawing plans, not overseeing construction. If you don't like the architect's plans, don't use them! Get new plans, or get a new architect.

WHAT MAKES A GOOD ARCHITECT?

Now you know how to behave as a client. But what makes a good architect? This is the $64,000 question (gets you a kitchen and a bathroom). Not every architect is good, and not every architect is appropriate for your project. There are several criteria that you ought to keep in mind when hunting for an architect.

Architects and Availability

Your architect will always be available at a price. But the materials they suggest may not be so easily found. Architects often just don't know how long something takes to order. Before you sign off on an architect's plan, discuss with the builder how quickly certain materials can be obtained. If it's going to take months, you may want to select alternative materials.

Architects As Contractors

As a rule, architects do not like to be general contractors. They do not enjoy overseeing renovation. Plans they enjoy. Talking with plumbers, electricians, and carpenters isn't so much fun for them. So the architect probably will visit your house from time to time but won't become involved in the everyday aspects of supervising.

However, your architect should establish himself as the general contractor's boss if you decide that this is the proper role for the architect. That is, the architect ought to be the one who guides the G.C. "I would have liked for the architect to have leaned on the builder a little more. We would have liked somebody who was tougher, someone who could have called the builder and said,

'You get your act together,' " Michele Sands says about her Ann Arbor, Michigan, renovation.

Design-Build Firms

These are the guys you may really be looking for when you think you want an architect. They solve problems instead of making them. Design-build companies occupy the middle ground between architects and builders. They offer one-stop shopping for the consumer. Design-build firms do it all; you don't have to worry about coordination between the architect and the builder, there's no running around to hunt for tile, no barking commands at the subcontractors, no anxiety over whether the drywaller will coordinate his work with the electrician (with a design-build firm, they've probably known each other for a long time).

Mark Richardson, vice president of Case Design, a Washington, D.C., design-build firm, offers this description: "A design-build firm is like a supermarket; it's a place where you can get everything you need."

Design-build firms are for people who want a relatively low-risk project. According to Mark, with design-builds, "You're not creating a triangle of liabilities and responsibilities. There are only two parties involved, the client and the firm. The client isn't caught in the middle between an architect and a builder." This, says Mark, means that you don't run the risk of "additional expenses because of a lack of communication" between the parties. Miscommunication is one of the major causes of renovation nightmares; the better the communication, the better the work, all other things being equal.

The second advantage to using a design-build firm also stems from the fact that you're doing one-stop shopping. A design-build firm helps keep you on budget. This isn't the same as keeping your project low-cost, because there are expensive design-build firms. Rather, because a design-build firm is in the business of building, once the plans are drawn you will have a very precise idea of the project's costs. "An architect isn't in the business of building," Mark Richardson points out. "He doesn't know what things cost. A design-build firm does. A design-build firm, for example, will know that an Andersen window may be 10 percent less than a Pella window, or that the labor to install one window will be less than with another." And so forth. In short, with design-build firms you have a much greater chance of preventing Pentagon-style cost overruns.

But there are downsides to using a design-build firm, too. Design-build firms may be the supermarkets of renovation, but sometimes what you want can only be found at a gourmet store. With one-stop shopping and low risk you get fewer choices with respect to materials and fixtures. With an architect you always have an unfathomably large number of choices for cabinets, windows, knobs, wood floors, plumbing fixtures, and just about every other part that goes into your house (some of these parts may take three months to arrive). With a design-build firm, the range of choice is more limited, although still large, and generally available immediately. For example, a given architect might work with fifty cabinetmakers, while a design-build firm might specify only five cabinet companies from whose wares you can choose. Again, Mark Richardson explains: "A design-build firm is not going to spec out things that it's not familiar with. If you want to be able to choose from among fifty cabinet manufacturers, go the architect route. If your time to oversee the project is more limited, then you want a design-build firm."

Both designer-builders and architects admit that there's an inherent conflict of interest with design-build companies. A design-build firm wants to use products that give it the greatest possible profit. They know that buy low, sell high is the route to profits. These firms will steer you to the higher-profit products and designs. Unlike using an architect in conjunction with a builder, there's no system of checks and balances with a design-build firm. You won't be cheated, necessarily, but it helps to know how the economics of design-build firms work.

If you use a design-build firm, it's your responsibility to play the role of checker. Ask the design-build firm to break down the price of different aspects of the work for you. (Make sure that they are willing to do this at the onset; otherwise, when you ask them later, they may balk. If they're not willing to give you that right in the contract, choose another firm.) If a price for purchasing and installing windows seems high for example, make some calls yourself. In a way, this defeats the purpose of using a design-build firm—that is, having the firm do all the work and much of the worrying. But with home renovation, there's no such thing as leaving everything up to the builder. You are the ultimate overseer.

Still, design-build firms are good for people who are not good managers or organizers. Design-build firms know the correct order in which to do things and who should be doing it. (If you are a superb organizer, then you might be suited to being your own general contractor.) According to Mark Richardson, management skills are more important than understanding how two pieces of wood go together. "If you are a good organizer," he says, "I would not discourage you from trying to tackle it yourself."

Get Details into Plans

No matter who you work with—an architect, a design-build firm, or just a builder—your plans should be rich in detail. This isn't just a practical measure; it could prevent you from falling victim to unscrupulous contractors. Some builders guarantee themselves winning a contract for renovation or construction houses because they bid the exact cost of construction with no profit built in. The low bid wins. These builders do make a profit, however: The profit comes from the change orders that he knows will be coming once construction starts. Builders who make their money this way typically develop stunted plans, plans that look fine on paper, but when executed simply aren't sufficient to meet anyone's needs. While this practice is not common, it does happen. And it's a good reminder of how expensive change orders can be: The profit these builders expect to make is 30 percent above cost.

3 | Contracts

Keeping Contractors Honest

THINK OF ALL THOSE jokes about lawyers. Well, this is one instance where they're worth every penny. Think about it: Does it make sense to sign a $5,000, $20,000, $100,000, or $250,000 contract without knowing exactly what you are signing? Especially when you're signing a contract for a project in which there's a 100 percent chance that something will go wrong, a 75 percent chance that something grave will go wrong, a 25 percent chance that there will be a disaster of some kind, and a 5 percent chance that you'll eventually end up in court?*

One in five homeowners has a serious argument with their contractor. As many as 25 percent of home renovators consult a lawyer during the renovation.

Although there isn't any statistical information to support this hypothesis, I believe that those people who consult a lawyer before they sign their contract are less likely to consult a lawyer during their renovation.

Bold type on the boilerplate contract provided to architects by the American Institute of Architects (AIA) reads: THIS IS A BINDING LEGAL AGREEMENT. IF YOU DO NOT UNDERSTAND ANY PARTS OF IT, CONSULT COMPETENT LEGAL ADVICE.

These words are true for any and all written contracts you

*Anecdotal estimates.

might have with architects, builders, suppliers, contractors, and subcontractors. Believe them.

If you spend $200 or $300 of a lawyer's time going over your contract, or have him draw up the contract (which will cost more than $300 dollars), it will be money well spent. Some of the potential dangers that a lawyer might spot in your behalf are

Whether there's a specific mechanic's lien clause in the contract;

Whether the contractor or you determine when the work is completed; and

Whether you're stuck with the contractor no matter how shoddy the work is.

He will also be able to clarify

When payments are made;

Whether there's a warranty; and

Who is responsible for obtaining permits.

When it comes time to select a lawyer to examine your contract, one criterion is more important than any other: Has the attorney in question renovated a house? If the answer is no, then the attorney probably has little idea of the kind of problems that you might encounter and may not know how to examine the contract in these terms. Being good at contract law is not the same thing as being knowledgeable about construction contracts.

If you decide to write your own informal contract, have a lawyer look it over too. Although you might get the contractor to agree to everything you want, if there's any ambiguity in the contract, the ambiguity is interpreted against the person who writes the contract, which means that if there's a problem with the contract, the contractor could win should you duel it out in court.

Take at least twenty-four hours to review any contract before you sign it. You never know what a good night's sleep might reveal.

There are innumerable elements that can go into a renovation contract; consequently, there is no such thing as a standard contract. Don't believe your construction company when it says, "Don't worry, this is a standard renovation contract." Just the same as with wars—there's no such thing as a generic war.

Examine the contract thoroughly.

One of the purposes of this chapter is to tell you what to look out for in renovation contracts. So why, you ask, should I bother with the expense of a lawyer at all? Well, even though the advice in this chapter will walk you through most of the renovation contracts you will see, every contract is different. A good lawyer will also be able to interpret your contract with respect to the laws of your particular state and city. Your state's laws may afford you more legal protection (or less) than the contract offers, and this is something you should know. Also, when negotiating the contract with your contractor, it's helpful to be able to use your lawyer as a crutch: "My lawyer said . . ."

Which brings me to the next item on the list of dos and don'ts: Always negotiate your contracts. Never sign any contract with a contractor without negotiating *something.* The reason I can make this bold recommendation without having seen your particular contract is that contracts are always written to favor the person who wrote the contract. The contract is to the initiator's advantage, which means there's stuff in the contract that is not good for you.

What to negotiate? Essentially everything. That's what the rest of this chapter is about.

Get Everything in Writing

Everything. "One homeowner had a verbal agreement with a builder to build a room a certain size," Ted Fleming reports. "The homeowner got a room that was four square feet smaller than he thought it would be. The homeowner didn't put the room's size in writing; instead he had a handshake agreement."

Many oral contracts may be enforceable by law, but oral contracts are a mess. The whole of the understanding between you and your contractor should be in writing, and the contract should say so in words like this: "This contract forms the complete contract between the Parties. No oral modifications are accepted.

Any changes to the contract must be done in writing."

There. That's pretty clear. Then if you ask the contractor to put in three additional electrical outlets and move a window an additional six inches, and the contractor later presents you with a bill for an additional $700, you can say, "What? I'm not going to pay for that because you didn't tell me in advance in writing how much it would cost." For your protection, and to a lesser degree the contractor's, written agreements and changes are the only safe course.

Get a Real Contract

A "real" contract is not a one-page proposal written by the builder, contractor, or architect. Signing such a document legally binds you to its terms, and its terms are deliberately unspecific—something that's in the contractor's favor. One-page agreements offer little or no protection for you, the homeowner.

Sign the contract in your home, not in the contractor's office. The Federal Trade Commission requires that the homeowner have three days in which to change his mind on any contract for more than twenty-five dollars if it was signed in an impromptu location, such as a home. Those extra three days may give you a little breathing room.

What should a contract contain? As I mentioned earlier, practically anything can go into a renovation contract, but at a minimum each contract should have these elements:

The full name and address of the contractor;

Your address;

The starting date and completion date;

Price and payment schedule;

Reference to dated specifications (drawings, detailed materials list, etc.);

How change orders will be handled;

A warranty on all work;

A nonassignment clause that prevents the contractor from giving the project to someone else, unless you both agree; and

Signatures and dates.

Beyond these elements, anything is possible. The rest of this chapter tells you what you ought to have in your contract to make it more secure, as well as what you should avoid having in your contract.

The more information in the contract, the better. There is no such thing as a too-detailed contract. In general terms, a remodeling contract is an equation, with your payment on one side and the contractor's work on the other side. The two sides should balance. The stuff that goes on your side is pretty straightforward: a dollar figure. The stuff that goes on the contractor's side is complicated. If anything is missing from the contractor's side of the equation, the contract is not detailed enough. For example, if you are having siding installed on your house, the contract should also specify gutters, window treatment, and eaves.

Although all contracts should have a warranty clause, the length of the guarantee period is obviously important. One year is a minimum; certain work such as foundations, framing, and ducts should have longer warranties. Will the warranty repairs be performed by the same incompetent plumber who did the shoddy work in the first place, or does the contract allow you to get another plumber? These are the kinds of points a good lawyer will bring to your attention.

Check your G.C.'s insurance. Does he have any? Your contract should specify that the G.C. will have all the necessary insurance as required by law.

Another nifty clause to include in your contract allows you to terminate—love that word—the contractor's services at any time if you are dissatified with the work.

For good measure, the phrase "Time being of the essence . . ." should appear somewhere in your contract. This puts the builder on notice that how long the project takes is important.

Watch Out for the Fill-in-the-Blank Clause

There are plenty of killer clauses in contractors' contracts. The most obvious requires that you turn over your house to the builder if you haven't paid him every penny he says you owe. (Don't laugh—it happens). But contractors also will try to include more subtle terms. One of the most popular has to do with digging foundations and says that if the builder encounters some unpredictable problem while excavating, he can charge more than is stipulated in the contract. Sounds reasonable, right? After all, if, while digging a grand hole for your addition, the builder finds a complete brontosaurus skeleton, he shouldn't have to pay for removing it undamaged and transporting it to the Museum of Natural History in New York. You get the idea.

But the reality is that foundation digging almost always runs into "problems": a sewer line that the contractor didn't know about (he should have consulted the city map), a layer of hard rock (he should have informed himself about the local geology), and so forth. Virginia homeowner Bob Adriance had this experience when his foundation was under construction. "They ran into problems and we had to pay about $2,000. In order to reach the line for the sewer, they had to dig deeper than they had thought; the county also required more gravel than they thought," Bob recalls. The most upsetting element of Bob's foundation complication was that it happened so quickly. No one wants to antagonize their builder at the outset—good impressions count, right? "It was at the beginning of our renovation," Bob says, "when we were trying to establish a good relationship." Another homeowner was snagged for $1200 because the builder ran into tree roots while cutting the foundation. Remember, foundation excavation problems are common and you should anticipate them.

There are two items you should insist upon in the excavation clause of your contract. The first is simple: Get the builder to put a cap on the cost. He'll be reluctant to do so because he doesn't want to scare you away with a high figure. He will say something like, "I really can't say; I could run into anything down there." Insist nonetheless, and say to him, "Well, think of the most likely problems and figure out the cost." This way, you won't be shocked when the bill comes due.

Second, you should limit any extra payment to problems that the builder could not reasonably have foreseen by consulting city maps, local regulations, geological surveys, or utility line listings.

Knowing about these things is the builder's job, and you shouldn't be penalized for his ignorance.

The Late-Penalty Clause

If you can get your contractor, builder, or subcontractor to agree to incorporate a late-penalty clause, that's terrific. Late-penalty clauses work like this: For every week (or other interval) beyond the contracted completion date that the work remains uncompleted, a specified number of dollars will be deducted from the contractor's fee. For example, if an addition is supposed to be completed June 1, the contract could specify that for every week work continues beyond June 1, $300 will be deducted from the final payment. Sounds good, doesn't it? And in fact, many contractors are willing to incorporate such a clause into their agreements.

But—and it's a big but—late-penalty clauses can backfire on homeowners, creating the most awful conditions.

One Seattle, Washington, homeowner got a late-penalty clause into a contract for an addition and general renovation. This homeowner had two bids for the project. The first bid was for $60,000, plus a 21 percent overhead fee (time-and-materials basis); the second was for a flat $62,000 (fixed-price contract). (The two basic kinds of contracts are time-and-materials and fixed-price. Each will be discussed in detail later, but briefly, a time-and-materials contract is a pay-as-you-go contract in which the homeowner pays for materials, plus an overhead percentage to the contractor. A fixed-price contract is just what it sounds like: a set cost for set work.) With the second bid, the homeowner was able to negotiate a late penalty of $1,000 per week—pretty advantageous from the homeowner's perspective. The homeowner opted for the fixed-contract bid with the penalty clause over the less predictable time and materials bid, a reasonable decision.

A late-penalty clause has to have a completion date to go with it, so the homeowner and contractor agreed on seven weeks from the start of the job—a target more or less pulled out of a hat. In retrospect, it was apparent that the project really was a four-month job, and therein rested the problem.

But a problem for whom? Most likely for the contractor, since the longer he took to complete the job, the less the homeowner would have to pay him. Or at least that's what the contract said.

Unfortunately, you can't legislate physics. When an immovable completion date meets an unresolvable complication, there's going to be a disaster. And that's what the Seattle homeowner got. At first the work proceeded smoothly. During the sixth week of construction the architect was invited back to inspect the progress. What the architect saw was "complete disregard for my plans." When he took stock, the architect says, "There was post-and-beam construction instead of foundation and load-bearing wall construction, as the plans specified. There were wholesale materials changes such as Masonite siding versus beveled pine siding." The homeowner, lacking building knowledge, was unaware of this. Even worse, according to the architect, if the builder had been allowed to continue, using the same substandard construction materials, the project would have taken an additional five weeks. Armed with the architect's report, the homeowner forced the general contractor to make design corrections, which alone took six weeks. Those six weeks added to the five weeks remaining on the construction of the project meant that the renovation was running eleven weeks behind. (A week of negotiations was also necessary to coax the general contractor into agreeing to make design corrections.) By this time, the general contractor had lost almost all profit on the job. The only thing he could do was to finish the job and hope that he made money on the next job. When a general contractor reaches this point, any incentives to do good or even adequate work are lost. The additional late penalties became meaningless to the builder, and the state of the homeowner's house became meaningless to him as well. Eventually, the contractor just stopped working, leaving the renovation and addition incomplete.

To make matters worse, subcontractors—whom the general contractor now could not afford to pay—placed liens on the homeowner's house. Liens totaling $21,000! These liens were in addition to the money the homeowner had paid the general contractor.

This is an example of how late-penalty clauses can work against you. If the general contractor has no way to make money on the project, he'll let it go to pot. Worse still, a late-penalty clause that backfires will make your project take longer.

The Work-Stoppage Clause

A work-stoppage clause can offer the right incentives, under certain circumstances. This clause allows you to use your remaining construction funds, after first legally notifying the original contractor, to hire a second contractor when delays become a stoppage. It's an invaluable clause to have because it allows you and your contractor to get a divorce, if necessary. You won't get out of paying what you owe the contractor, but you won't have to put up with him (or his subcontractors) any longer. Work-stoppage clauses are effective only if you have held back at least 10 or 15 percent of the total payment, and preferably 30 to 50 percent. I doubt if there are any contractors who are liable to voluntarily refund your money if you fire them.

Binding Arbitration

In short, binding arbitration obligates both homeowner and contractor to have disputes settled by a neutral party rather than by going to court. It forces both parties to continue talking rather than have lawyers do the talking. Arbitration is faster and less expensive than lawsuits, which can take years to come to court. If your contract has an arbitration clause, it should specify who will handle the arbitration; your architect, the American Arbitration Association, and the Better Business Bureau are three options.

Courts are always dicey propositions for homeowners, anyway. Even when you are in the right, you can't be assured of a favorable outcome. One former California judge put it this way: "[Homeowners] have so much money invested that everything has to be absolutely perfect. But the standard is that of reasonable performance. Nobody's perfect." In other words, you could lose in court, no matter how good your argument.

Leave Nothing Vague

When your contract says that you will make a *draw* (a payment after a certain period of time or after certain work is completed) after the installation of beams in the bedroom, does that mean "completed installation," "having them in place," or "installed but not yet sanded"? If the contractor is supposed to get three bids for each major subcontractor, does that mean he is supposed to let you choose from among those bids?

Spell out everything, so that there is no room for nasty arguments later.

Should Your Contractor Have Contracts with His Subcontractors?

More than likely, your contractor won't have subcontracts. For the most part, it's only important that you have a contract with the prime contractor. The downside of contractor-subcontractor contracts is that in the not-so-unlikely event that the contractor fails to pay the subcontractor, the contract helps establish proof for a lien on your house by the sub. Speaking in favor of a contract between the builder and sub is the greater firmness of a warranty. If something goes wrong, a written contract between the contractor and his subs helps the contractor get the subcontractor back to fix the problem. As a rule, it's a good idea to make sure that there are contracts with the more important subcontractors such as electrical, plumbing, and HVAC—the systems in which work is most likely to go wrong during a one-year warranty period (these areas usually represent less than 20 percent of the total cost of a renovation). In the absence of such contracts, the contractor might have to perform the repair work himself, something he's not necessarily qualified to do. With a subcontractor contract, even if the builder is out of business you can use the contract (make sure, of course, that you are given a copy) to bring back the subcontractor yourself.

More on the Issue of Subcontractors

Your contract with the builder should specify that you have the absolute right to approve the general contractor's choice of subcontractors. That is, no one whom you do not approve of can work in your house. If, during your investigations, you determine that the cabinetmaker the contractor wants to work with is lousy, no matter what the contractor says, you ought to be able to refuse that contractor. Without such a clause you can't. The contract should specify that when the G.C. gives you the bids from subs (if he's getting bids for a time-and-materials basis contract), he must give you the names of the contractors as well. You can then run a quick check by calling the Better Business Bureau, checking on bankruptcy, and asking for home-owner-references.

Try to Get a Specific Waiver of Mechanic's Lien Rights

It always stretches my credulity when I hear about mechanic's liens. In case you aren't familiar with how they work, a mechanic's lien is a form of partial ownership which may be placed on your house by a tradesman to whom the general contractor owes money. A mechanic's lien can be placed on your house even if you've paid your G.C. all his money. The mechanic's lien is something that Stalin might have invented. All that matters from the law's perspective is that the subcontractor hasn't been paid. Who hasn't paid him isn't an issue. A worker with whom you've never had a contract, never met, and never personally invited into your home can file a lien on your house, making you responsible for your contractor's debts.

A mechanic's lien doesn't mean that a plumber or electrician can take possession of your house, or move into your child's room. Though this isn't from lack of want. What it does mean is that until you get the lien removed by settling with the plumber, fighting the case out in court, or waiting for the lien to expire, you cannot sell your house and you probably won't be able to refinance your mortgage or get a home equity loan. While mechanic's liens rarely result in forfeiture of homes, they do encumber your house and limit you from selling it, refinancing your mortgage, getting a home equity loan, and all the other necessities of modern living.

Keep in mind that contractors, and materials suppliers too, can put a mechanic's lien on your house.

A waiver of mechanic's lien rights means that each subcontractor gives up the right to sue you for any money the G.C. might owe him on your job. By insisting on this clause in your contract with the builder, you give yourself added protection. A clause that says "the General Contractor will obtain a waiver of mechanic's lien rights from each subcontractor" may help you later. While the best way to avoid a lien is to make sure that the builder chooses reliable subs—and that your G.C. himself is reliable—the purpose of a contract is to anticipate the worst and be prepared for it.

A lien can be Excedrin Headache Number One. Eventually, most mechanics liens disappear—a lien must be followed within a certain time period by an actual suit for money or it will expire. Some subcontractors don't have the necessary evidence for their liens to survive a summary judgment, namely, proof that the subcontractor worked for the contractor, that the subcontractor is licensed, that the subcontractor was working with the correct permits, and that the work was being done with the knowledge of the homeowner (as opposed to a renter who might have had work done without a landlord's knowledge).

But some mechanic's liens persist, and in that interval they can wreak all sorts of havoc.

A most effective precaution against a mechanic's lien is to hire a reputable general contractor. A good G.C. will work hard to ensure that all his subs are paid; reputation is important for him. The second precaution is to make the contractor produce papers confirming that all subcontractors have been paid in full before you pay him in full. Simultaneously, you should get a waiver of mechanic's lien from each of the subcontractors before final payment is made to the contractor. You should ask your G.C. to get a waiver of mechanic's lien from all his subs before he makes the final payment.

In fact, try to get a waiver of mechanic's lien from each subcontractor before the work starts. By doing this, each subcontractor agrees that if the G.C. owes them money, they will go after the contractor, not your house. This is a difficult document to obtain, but it's worth asking for because some subs may sign it. The more who do, the safer your property will be. Not all subcontractors will sign such an agreement, though in economically tough times many will in order to get the job.

This document certifies that all the subcontractors have been paid: If the G.C. can't get his subs to sign a waiver of mechanic's lien indicating that they have been paid to date, don't give the contractor his final payment. You are at risk of having a mechanic's lien filed against your property. You should also insist on absolute proof that the G.C. has paid his materials suppliers, who can also file liens against your property.

Here's a sample waiver for a contractor-subcontractor agreement (courtesy Paul Locher of Locher Design-Build):

MECHANIC'S LIEN RELEASE

The undersigned Contractor hereby certifies this _____ day of _____ 19_____ that he has been paid in full for all work performed on and materials supplied to, the premises _____ . Contractor hereby releases Owner from any claims, liens, or any other court proceedings intended to secure payments in connection with the work performed by Contractor on the premises.

Contractor further certifies that he has paid his materials supplier in full for all materials delivered to, or utilized on, the premises. Contractor further agrees to hold harmless, protect, and defend Owner from any claims, liens, or any other court proceedings made by Contractor's materials supplier for supplies delivered to or utilized on the premises.

_____
Contractor

CERTIFICATION OF SUBCONTRACTOR PAYMENTS (HOMEOWNER TO SUBCONTRACTOR)

I, Joe Contractor, hereby certify that all subcontractors whom I hired to work on the house owned by Jim and Nancy Montana at

123 Sycamore Street, Aspen Ledges, Maine, have been paid in full and completely. I further certify that I have no outstanding monetary obligations to any subcontractors who worked on this house. I attest that a waiver of mechanic's lien has been obtained by me and signed by each of the subcontractors whom I employed on this project. I have received $10,354.00 as final payment for work on this house in consideration of these abovementioned assurances. These statements are truthful.

_____

Joe Contractor, GC

_____

Date

_____

Witnessed by (print name and sign)

I, Joe Subcontractor, hereby waive any and all rights to place a mechanic's lien on the property owned by Jim and Nancy Montana at 123 Sycamore Street, Aspen Ledges, Maine. I certify that I have been paid in full by Joe Contractor, through whom I was employed, for work I performed on this house.

_____

Joe Subcontractor

_____

Date

If a lien is filed against your home, the document above will make it easier to get the lien dismissed.

Liens become a dramatic headache if you live in a co-op building. A single apartment owner against whom a lien is filed places the entire building at risk. A 1989 decision by the New York State Supreme Court *(Dash Contracting Corporation v. Slater)* concerned a New York City co-op owner who refused to pay $21,380 of a $96,500 renovation project. Rather than going after the Slaters, Dash Contracting filed a lien against the building and the land owned by the co-op. Despite arguments that the co-op corporation was not involved, as it was not a party to the contract, the Court ruled that the Slaters were proprietary lessees and shareowners. In addition, the co-op board approved Dash's plans, legally involving them in the renovation. Further, the co-op as a whole benefitted from the Slaters' renovation. The judge concluded that because mechanic's liens can be filed only against real property, which the entire co-op owns, the co-op corporation is obligated to pay the debt. As a consequence, this co-op building's title is now clouded and the lien could prevent refinancing the mortgage, as well as shareholders selling their apartments. While this decision affects only New York state, it could be applied elsewhere.

Liens can strike anyone. When the housing department of the District of Columbia tried to sell six rehabilitated houses in 1982, they were prevented from doing so by $52,000 of outstanding mechanic's liens. "We screwed up," said the director of the Neighborhood Improvement Administration.

If you haven't taken these precautions, ultimately it's easier to settle with the subcontractor than go to court. Your first step should be to try and wring the money out of your contractor. Be as firm or as threatening as you know how. Get mean. Get inventive. Promise to make sure that he never works in your town again, that for all future times everyone will know that he's a deadbeat, no matter what company name he uses—and you'll make sure of that. Still, despite your attempts to imitate an organized crime enforcer, you probably won't be able to get a recalcitrant contractor to pay the subs who worked on your house.

If you've held off paying the contractor his last third of the payment, as you should have done, you will have a reserve from which to pay the neglected subcontractors. That's what happened to Bob Adriance, who remodeled a kitchen in his McLean, Virginia, house. "The contractor went bankrupt just after he finished the work," Bob says. The plumber whom the contractor had hired was due $5,000. "The plumber was going to put a lien on the house because he was owed the money by the contractor. I couldn't imagine that could happen. But I talked to a lawyer and he said the plumber could do it.

"Since I still owed $2,500 to the builder, his last payment, I told the plumber that I would pay him $2,500 of the $5,000 the plumber claimed the contractor owed." Bob also discussed this with the contractor, who said it was OK with him. "I told the plumber that if he went to court, he wouldn't get all of his $5,000 anyway. The plumber accepted my offer." Bob notes, "As long as you owe someone some money, you have negotiating power."

If you pursue this route, make sure that you get properly signed documents saying that both the subcontractor and general contractor will not pursue any further claims against your property. The subcontractor's document should additionally say that he will not seek to collect additional money from the builder after you've settled with the sub. That would be getting paid twice, and no subcontractor should profit at your expense. While you're at it, check that there are no other subs who might file claims. Ask the builder point-blank about the issue, and get him to sign a statement attesting that no one remains to be paid.

Incentives

Earlier I talked about late-penalty clauses. Since there is a relationship between time and money, a good contract will also have clauses designed to encourage the contractor to come in under budget. Bonuses give general contractors incentive to get subs to do their work at a reasonable price; they also give the G.C. the incentive to obtain the best possible bid. Bonus clauses work best with time-and-materials contracts. Such a clause might state that for every dollar the project is under budget, the G.C. will retain fifty cents (as you do), and that for every dollar over budget, the G.C. will pay 50 cents to you, in the form of a deduction from his payment (you'll never in your entire life see a check from a contractor to a homeowner).

Bids and Materials

Your contract should specify that the G.C. will get three bids for each major component of the project. The contract should also specify which materials the general contractor will use and that all materials and workmanship will meet the specifications of the American Society for Testing of Materials (ASTM) handbook or some other recognized standard, as well as local ordinances.*

Put As Much Detail into Your Contract As You Can Think Of

A bad contract reads, "Contractor Joe will renovate the kitchen of homeowner Tony." A good contract reads:

> Contractor Joe will renovate the kitchen of homeowner Tony according to the plans drawn by architect Valerie, dated March 11, 1991. In addition, contractor Joe will use the following materials (type, color, and brand). . . . Contractor Joe will work according to the following schedule. . . . Contractor Joe will be paid according to the following schedule. . . . Contractor Joe will seek the approval of homeowner Tony before hiring any subcontractors or making any deviations from the plan. . . . Contractor Joe will haul away trash once a week . . .
> and so forth.

You simply cannot get too much information into a contract. Even the types of nails to be used in different locations should be specified. If the builder's contract doesn't have space for this kind of information, create an addendum, a list that goes with the contract that both you and the contractor will sign.

For instance, you might expect that the cabinets your contractor installs in the kitchen will have pulls. But unless your contract specifically mentions cabinets with pulls (or refers to a cabinet that comes with pulls), technically this can be considered a extra, for which an extra charge is made.

Contractors don't want details in their contracts. First, details mean there is more for them to read and write. Second, details mean that they will have to do some things that might not turn out to be the easiest things (for them) to do. "There was a lot of

---

*ASTM's handbook is a good reference, but it will drive you (and your contractor) crazy with its level of detail. It's very difficult for anyone to do work 100 percent in accordance with the ASTM manual. The manual will say, for example, that the concrete should meet such rigid specifications that you would need to go to the plant to check the batch to certify that it conforms to ASTM standards. While it's good to throw in a catch-all clause that everything must conform with ASTM standards, don't hold your contractor to every detail.

pressure from the builder and the subcontractors," Michele Sands, a veteran home renovator, says. "They said, 'This contract is so picky and so full of details. You know, we've done this so many times.' They emphasize that they know what to do, and that their experience is better than a contract full of specifications. Wrong. Put in the details, because if it's not on paper, you may never see it in your house."

Never Pay a Subcontractor Yourself

Only the contractor should be responsible for paying subs. There's a simple and essential reason for this. The moment you pay a subcontractor, depending on your state's laws, you may be legally liable in case of any accident the subcontractor has on the job. When you write a check, you become an employer. Most homeowners don't have insurance to cover this kind of fiasco. So even if the subcontractor begs you, insist that the money come from the general contractor.

N. I. C.

N.I.C. means "not in the contract." Anything that's not in the contract—or shown in the drawings or listed in the specifications—is N.I.C. You can be charged extra for N.I.C. work—maybe. If the work was performed on a time-and-materials basis, you are responsible for paying the contractor. If the work was performed on a fixed-price basis and the builder didn't put the change in writing, when he comes to you later saying, "That'll be $700 for the door," you can fight about it. But it is much better to get in writing exactly what N.I.C. changes are going to cost.

Cleaning

Anything can be included in a contract. Some home renovators with allergies or asthma require the builder to clean and vacuum every day—thoroughly. But you don't have to have a medical problem to insist on a daily cleaning clause in your contract. As I mentioned in Chapter 1, "Surviving Your Renovation," daily cleaning will make the final cleanup not only easier but possible. Without daily vacuuming, sweeping, and even a little straightening up, you may never rid your house of all the dust that has been created. Make regular cleaning a component of your contract if you want it. When cleaning is not in the contract, you won't see any going on.

Cheap workmanship is like cheap wine. The price is right, but you regret it later.

L. L. Chambers, stonemason

4 | How Much Should You Pay?
(And Whom Should You Pay Off?)

LIKE SNOWFLAKES AND PREGNANCIES, no two renovations are alike. This is true for practically every aspect of remodeling, including how much it's going to cost. While there are formulas to estimate how much a renovation should cost—x number of dollars per square foot for a particular part of the country—a builder can't give you even a rough estimate without first looking at your plans and your house.

The variables that go into determining how much a particular project is going to cost also play a role in determining the *risk factor* of that project. The risk factor is a loose indicator of what can go wrong: cost overruns, bottlenecks that result from unknowns in your house, labor shortages, or the work falling apart. The variables that comprise the risk factor include

The complexity (including the age) of the house or apartment;

The complexity of the site (how access is gained, for example);

The experience, temperament, and contentiousness of your general contractor and subcontractors;

The finances of your contractor;

The schedule you've set;

Your financing constraints;

The special uses for the construction you're planning; and

68

Out-of-the-ordinary construction you want done.

If any item on this list looks high—that is, makes you want to reach for the Pepto-Bismol—your project is riskier than usual, and you can expect some sort of problem. That problem, when it occurs—and it will, because no renovation is without risk—will cost you money. The persistence and universality of risk means that you must expect to pay more money than you anticipated.

The risk factor remains with you throughout your renovation. Each component of the project carries distinct and unpredictable risks. In other words, risk never disappears; it is simply transferred from one stage to another.

The implications of the risk factor is that as you're budgeting a renovation, you shouldn't cut costs close. Plan for the renovation to cost more than you expect. For example, let's say you're adding a family room off the kitchen. The architect-builder tells you that the project will cost $35,000, so you take out a home equity loan for that amount. Expect to dip into your savings, probably to the tune of $5,000, for the project. It's likely that your plans didn't include everything you actually want for the room: painting, carpeting, lighting, and so forth. There's bound to be something you left out, and that something takes money. Similarly, you've got to furnish the room, so consider the cost of furnishings a renovation cost. While it's true that you can always furnish a room later (even years later), you won't get much use out of the room unless it has furniture. Then there are unexpected costs. Even if you are working with a fixed-price contract, unexpected costs pop up. As the project progresses, you might decide that you want a different window, a different built-in cabinet, a different flooring. Or the worst could happen: The construction disassembles; the contractor goes broke; subcontractors file liens on your house; and you end up paying for the same work twice.

A wise sage once said,

Have available a 15 percent contingency fund.

This is especially true once the project has passed the planning stage, when the responsibility for cost containment goes to the general contractor (in varying degrees, depending on the type of contract). You no longer have control of the way money is spent. While the builder may be more experienced than you in curtailing costs, it's your money (and house), and no builder cares as much about your house or your bank account as you do.

What Costs So Much, Anyway?

You want to remodel the second floor of your house: enlarge the master bedroom, enclose a screened porch, and turn the bathroom into a palace. The company you've hired quotes an estimate of $160,000. Where does this money go? Obviously, much of it ends up in the hands of manufacturers of sinks, whirlpool tubs, lamps, lumber supplies, and other materials you are using. However, in a typical renovation, some 60 percent of the money you pay can go toward hiring subcontractors, workers who are not full-time employees of the company you've hired! Following money often leads to revealing information. When you hire a general contracting company, you are not hiring their technical building skills per se. You are hiring their judgment to hire other workers and to oversee those workers.

Making Decisions

Throughout the project you will be called on to decide things. This is absolutely true for a time-and-materials project and somewhat true under a fixed-price contract. You'll have to make these kind of decisions: Do you want the faucet placed here—or here? Do you want the door to open this way or that way? Many of these decisions will be important, so you will want to take time in making them. If your contractor says he needs to know by tomorrow, however, decide by tomorrow. Frequently faster decisions mean lower costs because your contractor may be able to acquire materials at a good price—today only—or your contractor may know a reliable subcontractor who wants to work right away (but not next week).

Fixed-Price Versus Time-and-Materials Contracts

Generally speaking, there are two types of contracts: fixed-price or time-and-materials. Fixed-price is the simplest type of contract, so let's look at that first. A fixed-price construction contract is similar to the contract you invoke when you go to the doctor for your annual checkup: You'll get ten minutes talking with the doctor, ten minutes having your body pressed in uncomfortable ways, one X ray, one blood specimen, one urine sample—plus, you know exactly what the checkup is going to cost ahead of time. Basically, with a fixed-price contract, you show a builder the plans and the builder gives you a price. If you make no changes in these plans, the amount quoted is exactly what you end up paying. This type of contract is also called a lump-sum contract. You pick from among the lowest bids. Then if the cost of remodeling goes up, you're safe. Conversely, if the cost of remodeling goes down, the contractor saves. This is a classic zero-sum game. With a fixed-price contract, the incentives to the contractor to keep costs low are very high (as are the incentives to cut corners).

Under a fixed-price contract, when the costs to the contractor rise more than he expected, your savings aren't the only consequence: The contractor's willingness to do a superior job may vanish. Just as the price of the work is absolute, so is the builder's inclination to work on the project rigid. For the $40,000 you contracted for, you get $40,000 worth of work from the contractor. If the project unexpectedly costs $50,000, you still get $40,000 worth of work. Can you guess who ends up paying in the long run?

A fixed-price contract lets you avoid financial risk, but it doesn't eliminate the risk factor. Higher construction costs just transfer the risk from the area of financial hazard to quality control. With a fixed-price contract you have given the decision-making powers over the project to the contractor. He decides how to manage the expenditure of funds: where to spend more, where to spend less, where to reduce costs dramatically. A good plan may specify general budget constraints, but the actual costs of materials and labor and how they are skewed won't be decided by the plan.

Despite these drawbacks, a fixed-price contract is right for certain kinds of projects. Fixed-price contracts work well with small projects (small being a relative term), projects where you won't be making any changes as the work moves along, and projects where the construction parameters are well known. For a fixed-price contract to work, the builder must see the complete plans.

With a fixed-price contract, you don't have to monitor costs, *but you do have to monitor the quality of workmanship and materials* more than you would with a time-and-materials contract.

A time-and-materials, or cost-plus, contract is akin to visiting a lawyer or an auto mechanic: The number of hours the lawyer or mechanic works, plus the cost of the parts (or in the case of the lawyer, photocopies, paralegal's time, and telephone expense), plus a markup on those costs is what you will pay. You can limit your costs by talking quickly with the lawyer or by doing a patchup job at the mechanic's shop instead of replacing a part. Time-and-materials is offered when the contractor doesn't give you fixed price in advance for the work. What you pay is the cost of materials and labor, plus an hourly salary for the general contractor, plus a fixed percentage as markup. As opposed to a fixed-price contract, in which the contractor bears the risk for increased costs, with a time-and-materials contract you carry this risk. Worse still, the more the contractor spends on materials and labor, the more he makes (because of the percentage markup), so there is every incentive for the contractor to keep costs high. (Now you know why it takes so long to get your car repaired).

Despite these dangers (and there are ways to limit the risks through the contract, as you've seen in Chapter 3), a time-and-materials contract is good for certain homeowners. Only a time-and-materials contract allows you the flexibility to make major decisions as you progress. Where quality is important, time-and-materials contracts work best, since the incentive for the contractor to cut corners has been eliminated. Under certain circumstances, a contractor may not be able to determine from the plans the scope of the work, and a time-and-materials contract is the only type that is possible. Finally, a time-and-materials contract works well for homeowners who plan to supervise much of the remodeling themselves. A time-and-materials contract allows you to save money by putting your own skills and knowledge into the project. With a time-and-materials contract, for example, you can go out and purchase the tiles you want for your bathroom personally (and possibly still get the builder's discount).

You will pay a markup of between 15 and 25 percent to the builder on a time-and-materials contract. But the contractor can usually buy materials at a deep discount, so even a 25 percent markup may not be as large as it sounds. The markup should be a smaller percentage for more expensive projects, larger for smaller renovations.

Another advantage of a time-and-materials contract is that you don't need to put down a large deposit. You hold on to your money longer, but you have to be prepared to write checks more frequently.

You probably should have great trust in your builder with a time-and-materials contract. After all, what's to stop him from keeping the costs as high as possible? Beyond references, the chemistry between you and the G.C. needs to "feel right"; otherwise, a time-and-materials contract can turn against you.

Don't let a contractor's time-and-materials bid enter into your selection process. A bid is meaningless with this type of contract, since there's nothing holding the builder to the bid. You need to monitor costs closely with this kind of contract.

There are also variations on the time-and-materials theme. A time-and-materials/guaranteed-maximum-price contract gives the builder the flexibility to get the best materials and work (which you pay dearly for), but puts a cap on the project. If the builder spends too much, he loses; if he spends under the maximum, he reaps the profit. A time-and-materials/guaranteed maximum-price contract shares risk. You and the builder are part of the same team, and while your goals aren't identical, you have a commonality of interests regarding costs. The contractor's profit is a product of the quality of work.

It's important to talk to a builder's references before agreeing to a guaranteed-maximum-price contract. Did he remain within the boundaries of the projected costs on other jobs? Ultimately, however, you probably need to monitor costs less with a guaranteed-maximum-price contract than with a time-and-materials contract.

The differences between these three kinds of contracts are schematic. A little game theory went into describing the builder's incentives and your risks for each. Bear in mind that, for you, the relative merits of these three types of contract depend highly on the builder's temperament and experience, your wants, and the nature of the project.

The Bidding Process

Find three qualified builders-architects-design-build firms (having bids from more than three is often counter-productive). Give the builders adequate time to prepare their bids. You may decide to open bids in the presence of bidders at a stipulated time and place, if you think this will let the builders know that you are going to take their bids very seriously and that there will be real competition. Your inclination should be to award the job to the lowest bidder, but if the lowest bid is too low, take care. If a builder is losing money on a job, the work may suffer. As your parents taught you, you get what you pay for. Too-low bids send up red flags. Low bids may also be a signal that the builder is going to shortcut some legal requirements such as hiring undocumented workers, foregoing permits, using substandard materials, or pressuring you to take overstocked materials.

The more detailed your plan, the more precise the bids will be and the less risky your renovation will be.

Writing Checks

Periodically, you will have to write checks. Make progress payments up to a maximum of 90 percent of the work performed even after the work is completed. To protect you from mechanics leins pay the remaining 10 percent only after the builder has given you evidence that he has paid all his bills.

The project is going to cost more than you expected, even with a fixed-price contract. Keep a 10 to 25 percent contingency fund available to pay for it. Some remodeling veterans swear by the Rule of Three, which states that everything costs three times more than you expected (and takes three times longer). What follows is a list of some of the human and technical reasons for cost overruns. Pay attention to this list, because it's also a list of ways to plan to avoid cost overruns.

Contractors who go bankrupt, forcing you to find someone else to finish the project

Contractors you fire, forcing you to find someone else to finish the project

Contractors who try to convince you to alter your plans to use more expensive materials

Your decisions to change the plans midstream

Extra excavation

Poor soil

Larger footing

A change in the specifications of the roofing material

Popped walls

Shimming out crooked existing walls and drywalling

Specially cut materials to fit crooked walls

Replacing subflooring to support a tile floor

By the way, if you have a credit card tied in with a frequent-flyer program, use that card to charge as many supplies as you can. There's no doubt that you'll earn more than enough points to take a trip to the Caribbean, which you will need when the work is completed. (One of life's ironies: Many homeowners want to take a vacation after their renovation is complete rather than spend time in their house.)

No matter what you do, no matter how much you are paying your contractor—or how little—do not make the final payment until the work is totally finished. And checked. Notice that this is not a "you should" but an imperative: Do not! As the singsong goes, if you pay your contractor before 100 percent of the work is completed, *you'll be sorry.* One New Jersey home renovator didn't follow this advice and tried to sue his builder to recover the money. Court papers were served against the builder, who was by then living at the Kough Dwyer Correctional Facility.

But don't just take my word for it, ask Craig Stoltz, who discovered, "The Cayman Islands were the ultimate destination of my contractor."

"We had been doing about a $15,000 job on the kitchen," Craig laments. "We got to the final details of the project, the installation of the dishwasher and the fluorescent lights. He had a subcontractor install the dishwasher, and he himself installed the lights. But the dishwasher wasn't getting any water. There was a kink in the pipe or something, a fundamental problem. And to get the lights to go on, you had to flick the switch and then rap the lights.

"He asked for his final payment and we said, 'No, fix the problems first.' At this point we owed him one-third of the total payment. He said, 'There are only two little problems. I'll have somebody come and fix them. Just pay me.' We again said, 'No.' Then he went into a controlled rage, said that these really weren't problems, that they were 'bull.' But we held firm. Finally, we did say, 'We'll pay all but $200.' But he continued with his tirade because we wouldn't give him the full amount. Eventually he left our house and said, 'Pay me whatever's fair,' which was everything but $200. He took the money, went away, and never came back—because, we found out, he had gotten a job in the Cayman Islands. He had had no intention ever of completing the work."

Saving by Doing Some of the Work Yourself

Roughly 40 to 50 percent of remodeling costs are for labor and 23 percent are for overhead and profit, according to Bryan Patchan, executive director of the National Association of Home Builder's Remodelers Council*. That leaves 27 to 37 percent for materials. Clearly, you can save by doing some work on your own.

There are several questions you should ask before taking over part of the renovation yourself:

First, Can you do the work? Are you qualified?

Second, Is it worth the savings? Generally, a professional can do the job better and several times faster than you can.

*Banks, William C., "Knowing When to Do It Yourself," *Money* (June 1988).

Third, Do you want to do this work because you think you can do a better job than the contractor? Do you think you will have more patience, skill, or quality control than the contractor?

Fourth, Will your work interfere with the contractor's work? Will you be able to mesh and coordinate your work with the contractor's schedule and technique?

This is not a book about how to do your own renovation. There are plenty of fine books and courses you can use to learn the home mechanics. Most people do some work, even if it's only painting one cabinet. Whatever you decide, let your contractor know beforehand. And don't forget to discuss receiving a discount because there's less work for him to do.

What Happens When Your Contractor Goes Broke?

Have you ever been audited? Or received a parking ticket? The chances of your contractor declaring bankruptcy or going out of business in the middle of your renovation is somewhere in between the odds for these two unpleasant events. If your local newspaper lists bankruptcies, you'll notice that many of the companies going out of business are small construction firms. If you follow that column long enough, you will eventually see the name of a company you've invited into your house.

The attrition rate for startup firms is frighteningly high. Nearly 90 percent of new contractors don't make it past their fifth-year anniversary. This doesn't mean you shouldn't use a new contractor. Often someone who has just started a home renovation business has more enthusiasm, energy, and time to devote to overseeing your project. Just make sure that the investigation you perform on any construction company is doubly comprehensive for a young firm.

Check business references going back five years. At a minimum, ask your contractor for the names of subcontractors he's done business with. Call them and ask, "Has Mr. X paid you on time? Has Mr. X paid you the full amount you subcontracted for?" If the answers indicate that the contractor you're considering hiring has trouble meeting his obligations, don't hire him. Depending on the size of your renovation, you may want to dig a little further. Ask your contractor whether he has ever done business under other names and what happened to these businesses. Then check out his answers. Hire a credit-checking service. You may discover that the contractor has had several construction companies that have gone bankrupt; each time one goes belly-up, he forms another. If so, this is not the contractor you want to hire.

Even if you've done all your homework, there's still a chance that your contractor will disappear one day, sort of vanish in the Bermuda Basement. Although your renovation may be proceeding according to budget, your contractor could be facing a crisis on another house that puts him out of business.

Now what? Well, you're going to have to find another contractor. Equally pressing is the fact that the contractor's subcontractors will want to be paid—and they're going to come to you for their money.

Legally, you owe these subcontractors money. What can they do to you if you don't pay? The subcontractor can take the contract he signed with your contractor to court and get a mechanic's lien on your house (see page 58).

Certainly, you want to find another G.C. if the builder goes kaput during your renovation. Just be aware that because the first G.C.'s subs may never have been paid, if you hire a new contractor you may end up paying for the same work twice. Before you hire a new builder, settle things with the subcontractors who were working on your house.

Dealing with Dishonesty

It's a good idea not to become the prey of contractor-crooks, who rob Americans of tens of millions of dollars a year. Like weeds in summer, contractors pop up. They promise shining additions, new roofs, clean fireplaces, high-efficiency heating systems—then take your money and produce junk or nothing at all. Dishonesty is a virulent, contagious disease among contractors. One Fairfax, Virginia, contractor accepted an $8,000 deposit for a $54,000 addition, then vanished to contractor heaven. Never hire someone without investigating him, paying particular attention to his assets. Pay as little up front as possible. Use only licensed contractors, because when you use a licensed contractor the state or county offers you certain protections. Always, always listen to your instincts when it comes to money.

What can you do if a contractor disappears with your money? Do what a Fairfax family did: They called the police and had him arrested for obtaining money under false pretenses.

EVENTUALLY YOU'LL PAY, IN ONE WAY OR ANOTHER

There is a story about a builder who swindled his mason by not paying him for installing the brick veneer on four houses—the beginning of a chain reaction. The mason apparently had sub-contracted the masonry work and didn't pay *his* subcontractor. The subcontractor's subcontractor decided to retaliate in the form of a twenty-four-pound sledgehammer smashed against the brick veneer around the perimeter of the four houses. The brick, of course, collapsed from the facades. This would have been a disaster for the homeowners, except for the fact that the vengeful subcontractor's sub destroyed the brick on four houses that were not the four houses he had worked on. He'll be out on parole soon, so let the targeted homeowner be warned.

The Sales Commission

Whether this information is useful is up to you, but you should be aware that some sales representatives for design-build firms earn their salaries by commission. In some areas of the country these commissions exceed 12 percent. Here's the math: On a $40,000 renovation, the sales rep may receive $4,800. That's a hefty incentive to convince you that his firm is the right one for your job, and the sales rep will shine. I believe that contracting companies should disclose whether their reps work on commission. No company will tell you this up front, so you have to ask. You should not deal with any company that doesn't answer this question in a straightforward manner. If the company's sales rep does work on commission, then it's a wise idea to meet the people who will actually be doing the work on your house—the ones without the ties and jackets.

Expect to Pay

You are responsible for your debts. Unless there is a problem with the work, you must pay your contractor. You can't decide toward the end of the project that the renovation was too expensive and resolve only to compensate the builder for what you think the work is worth. Many homeowners do pursue this strategy in creative ways, but the ways end up backfiring.

Take the Aspen Hills, New Jersey, couple who tried to cheat their contractor. I was having dinner in New York City with my parents, talking about this book and about sex and contractors in particular, when Vincent Menonna at the table next to ours politely interrupted. "Women have sex with their contractors all the time," he said. Vincent is a licensed electrical contractor. We put our salads aside and listened. The Aspen Hills couple had just completed a costly renovation, but had yet to pay the final $60,000 to their G.C. During the renovation the wife-homeowner, with her husband's blessing, had had a roll in the hay with their contractor. A well-photographed romp, it's important to add. When the contractor asked for his final payment—no insignificant amount—the couple said that if he insisted on being paid they would present the X-rated photographs to the contractor's wife. Nice, neat blackmail, and probably also a potential entry in *The Guinness Book of World Records* for most expensive sex. However, the contractor's wife continued to nag her husband about the tens of thousands of dollars he was still owed. Eventually the contractor's wife had enough of his stalling and called the woman who owned the house. The homeowner-wife told the contractor's wife, "Your husband has already been paid. . . ." Now the contractor's wife knew, so there was nothing to lose by demanding the remaining $60,000, which the contractor quickly and successfully did.

Pay Attention To the Warning Signs of a Company Going Sour

A construction firm isn't going to go bankrupt just before you start doing business with them. Problems occur only when a contractor is in the midst of working for someone. You.

Having a contractor go bankrupt in the middle of your renovation is among the worst things that can happen: When a contractor goes bankrupt, it usually means that you aren't going to get your money back and you aren't going to get the work done (and possibly unpaid subcontractors and materials suppliers will file mechanic's liens against your home). It's an unrighteous disaster. It's best to be constantly vigilant against financial unsoundness.

Large size in a firm doesn't guarantee financial solvency. Nothing does, in fact. A low overhead helps, if you can determine whether or not a company has a low overhead, but this, also, is no assurance. Some remodeling experts say that a company with large assets gives you good protection. "Another reason we went with our firm was that they had an investor," Michele Sands says. "They had a backer, lots of capital. No problem of the company going under."

Her husband, Harry, adds, "With the other companies, we even talked with the architect about getting a completion bond. Here's this little company, it's only five guys, all one family. What happens if they go under? That wasn't a concern with the company we picked. They have big offices, they have all these projects, twelve crews . . . "

"But," Michele interjects, "the investor and the company had a falling-out. We don't know what it was about, but the investor pulled out."

The first sign of financial trouble is a contractor who starts late. If the work has been promised to begin on a particular date and the contractor says he's running behind on other work, or if there is some personal problem that's delaying the start, or if the materials aren't ready, the translation is: He has money problems.

Believe it or not, another sign can be tardiness in presenting you with bills (or, of course, unusual swiftness). Often the first person to be fired from a financially troubled remodeling company is the bookkeeper, the person who prepares the invoices. As a result, it may take a while for the company's owners to figure out who owes what and prepare the bills.

If someone you've worked with at the company abruptly leaves, there's a good possibility that that person is leaving because he's not being paid. "When George [the sales representative] left the company," Michele Sands says in retrospect, "that was the first sign."

When a company appears to be spending an excessive amount of time or using an above-average number of man-hours to accomplish a task, this indicates there is danger ahead. The company that dug the hole for Michele and Harry's bedroom addition took months instead of weeks and used all sorts of machinery week after week. They were clearly over budget. By the time the hole was completed, $50,000 of a total $75,000 contract had been spent. Going over budget in the early stages of a project means money troubles ahead. Sometimes a construction firm can absorb these temporary losses, but sometimes it can't.

If the company asks for money ahead of schedule, this too is a warning sign, a dangerous sign. "One morning," says Michele, "the president of the company showed up at our front door and wanted a check for $5,000 for a draw. It was due the next week, but he wanted the money then. He had a desperate look on his face. He looked exhausted. It was just weird."

Michele continues. "In March [1989] we realized we had to reorganize everything or we were really going to be screwed up. We started to think about trying to find someone else to finish the project. We were trying to figure out how to handle money."

Michele and Harry renegotiated their contract in the middle of the renovation. They were able to accomplish this because they threatened to drop the company entirely—and the company knew that it was having difficulties. "The contracting firm had to get a new crew immediately. A new foreman whom we liked had to be here full time," Michele says. "And we were to start paying the subcontractors directly because of the company's problem with payroll. We decided that we wanted to completely control the flow of funds. I felt much, much better. We get these check requests and we write a check to the lumberyard and it's entered against our contract amount. We're not going to go over the contract amount. We shouldn't have to do it at all." (Paying subs yourself isn't a good idea; but sometimes you have no alternative.)

If you do have to take over a project, try to get your money's worth. Don't count on suing the contractor later to get back *x* number of dollars for uncompleted work because chances are you won't be able to. Either it will cost you too much in legal fees, or the company will have gone bankrupt, or the company will have changed names and you won't be able to find them. If you're like Penny and Don Moser, you'll find yourselves as one of ninety claimants against the company. "Our coclaimants included Chase Manhattan, so we figured our chances weren't good," Penny says. Instead, squeeze every ounce of work out of the contractor. Whatever they've done wrong, get them to fix it, no matter how long it takes. Have as many materials delivered to your house as you have room for. Arrange for as much preparatory work to be done as they can do. Then fire them.

If your contractor is doing bad work, terribly tardy work, or has other major problems, and then, miraculously, things start getting better, don't be awed by this change. Don't assume that the project is going to go well from now on. This is just *contractor remission,* a temporary manifestation. When work goes from bad to good, it will go back to bad. Regulate your payments accordingly. Don't whip out the checkbook and write another check for $10,000 if last week you were about to fire the company.

Local Checking

Most workers need money right away in order to buy supplies or pay their bar bills. If you pay your workers with a money market account that's linked to an out-of-state bank, you may not get them to show up again. However, you probably can get away with paying a building firm with an out-of-state account and enjoy the few day's float. If you want to be nice, ask the builder if the out-of-state account is okay.

A Good Work Environment

A good work environment makes the work go faster, makes the work go better, and consequently makes the work cost less. When you give people a poor working environment, expect that the work is going to take longer.

What makes a good working environment? At a minimum, a telephone and a toilet. A warm or a cool place to rest, depending on the season, are good second starts. Chairs and an on-site radio

84

(one you're willing to lose) aren't bad ideas either. Making the work site as empty as possible is a big help. A space large enough to walk through helps too. Don't forget a place where the workers can store their materials: They like that. A working refrigerator can go a long way toward keeping the workers refreshed, especially if you stock it yourself with sodas. Consider a coffee machine. Keep pets or small children away. Occasional pleasant surprises also help the atmosphere; once, during our remodeling, we sent out for pizza for the whole crew, and they appreciated it.

Whom Should You Pay Off?

Sometimes the parade of city inspectors checking over your work makes you feel like you're crossing an international border: You can almost hear the vibrating tingle that emanates from the inspectors' palms. The custom of bribing is long and venerable in construction, yet bribing predominates mostly in commercial building and renovation, where it often yields dangerous, substandard work. My philosophy on this score is, Don't pay anybody in cash for anything. That solves the problem of whether to pay off someone or not.

One exception to the no-cash rule: It's okay to tip a subcontractor or buy him a present if the work was exemplary.

5 | The Permit Process

(Or, Does the City Like the Color of Your Toilet?)

WELCOME TO A NETHERWORLD where Rod Serling writes the scripts, Rube Goldberg builds the sets, and Franz Kafka casts the actors. This is the land of Permits and Variances.

Builder Paul Locher was doing a bathroom and study in a cooperative apartment building on Massachusetts Avenue in Washington, D.C. The unit was on the fifth floor overlooking the Japanese Embassy, not a bad view, because Massachusetts Avenue, also known as embassy row, has on it some gorgeous stately mansions. Paul had to get Fine Arts Commission* approval for the bathroom interior because the permit center felt that the room would be visible through the windows from the Japanese Embassy grounds, itself still under construction. This wasn't a typical bathroom renovation: The couple was planning to spend $50,000, yet they were required to present their plans to the Fine Arts Commission. Granted, this is an extreme example (the United States Fine Arts Commission only affects projects bordering on Federal lands, mostly in Washington). Still, it's an example of the kind of unexpected bureaucratic hurdles you are likely to encounter. In this case, the Fine Arts Commission ruled on the tile colors, the plumbing fixtures, and the relationship of the

---

*The Fine Arts Commission is a body appointed by the president that maintains aesthetic standards on our national sites.

fixtures to the windows. (Never mind the commission's implied but unstated assumption was that the owner of the apartment would leave his bathroom window shade open most of the time.) The commission's decisions cost the apartment owners thousands of additional dollars, not to mention weeks of delay trying to cut through steel-reinforced red tape.

Why Permits?

Permits are an outgrowth of government's efforts to 1) ensure that all construction is done safely, 2) exert quality control over materials, and 3) have a standard system of materials and construction methods.

It's the last element that sometimes rubs homeowners the wrong way. Standardization, like anything that conforms to an average, may not be what you want or appropriate for your home. For example, many cities require that sink faucets be a certain height above the sink. This "standard" prevents homeowner suffering from plumbers who are lazy or no good and who install faucets too low. Low faucets are hard to use—but you may *want* a particular faucet set low because it looks spectacular in your bathroom.

Who decides what meets code? Five national building organizations and many other professional construction associations including electricians and plumbers have developed model standards that many cities and counties have either adopted directly or modified. These standards become codified into law. Variations occur from area to area because of the availability of supplies, local geology, climate variations, the way buildings are used, and local policy decisions. For example, new houses in Maine are required to have more attic insulation than houses in Georgia.

It's a rare contractor who does not obtain permits. Homeowners who do their own work, however, often venture without permits, obtaining one only if a neighbor rats on them. There are a number of drawbacks to ignoring the required permits, not least of which is that in the event of a calamity—a fire, a flood, a collapsed wall—your insurance company may refuse to pay. Other drawbacks include fines if you're caught and reduced value of your home when you decide to sell (work not up to code almost always translates into a lower resale value). On the other hand, some locales will not allow unlicensed homeowners to get a permit for certain types of work, putting them in a damned-if-you-do, damned-if-you-don't situation.

While rare, a few contractors will work without any permits. A larger percentage will do work without some permits. This happens most often when one tradesman performs the work that another was supposed to do. Beth Farker's carpenter was also her general contractor. And painter. He was not supposed to be her plumber and in fact told Beth, "I have an experienced plumber I'll be working with." But as Beth told me, "I never saw the plumber, but I saw the electrician. I called him the phantom plumber. I found out that he did his own plumbing, though there was a plumber he consulted with. The plumbing turned out to be the least satisfactory part of the job. I have two sinks and a dishwasher. The stuff backed up into one sink rather than going down the drain." Although the carpenter eventually fixed the drain, and although the problem wasn't high on the Adler Scale of Disasters, this was not something that Beth relished. The origins of the problem? The carpenter didn't apply for a permit, so he could get away with not hiring a licensed plumber.

Hand in hand with permits goes inspection. The inspection process usually requires that your renovation pause in several places, often the most inconvenient ones. After an initial permit is obtained for electrical work, for example, an inspector will have to examine the work before a wall is closed in around the wiring, which means that the drywaller will have to wait until after the inspection. Waiting for an inspection can take days or weeks, depending on how busy the inspectors are in your area. If you run afoul of your local inspector, there's a good chance that he will issue a stop-work order.

If the work you're planning entails inspections, make sure that the general contractor is aware of the required inspections. Most of the time, this won't be a problem. What can be a problem, however, is a situation when there is no general contractor or when you are the G.C. An electrical subcontractor may not care so much about how quickly the drywall gets completed, and so may not bother to arrange for an inspection. One way to avoid this problem (other than berating the subcontractor) is to contractually tie the subcontractor's final payment to a successful inspection, giving the subcontractor the necessary incentive to get the inspections completed.

The Build-and-Cover Method

Some general contractors use the "build-and-cover" method of construction. If this sounds shady, that's because it potentially is. Build-and-cover—which we will see has some practical uses on some small jobs—runs counter to the normal way of construction. It's employed by some general contractors to do quick and inexpensive work. Normally, subcontractors are used in a series. For example, the demolition people will arrive; followed by the carpenters; then by the mechanical system people, insulators, and drywallers; then the carpenter again for final trim details and *punch out* (the finishing touches). In the build-and-cover method, there are no separate subcontractors; a general contractor takes the initiative of hiring labor to perform all aspects of the work. The workmen will attack the job on a location-by-location basis rather than as an integral system. The contractor selects a wall, for instance, and does everything to that wall in one day: framing, window, insulation, electric lines, and drywalling. The workers then go home for some beers. They're back the next day (hunting season permitting) to do another wall. This is risky for the homeowner because there's no method of testing the whole system. For example, you can't inspect the electrical system as an interconnected unit or examine the framing for integrity. You won't know how many plugs are supplied by each electrical line, nor will you know if the insulation has been put in properly. Generally, the work is done so quickly—the only advantage to build-and-cover—that there's no time to inspect or correct the work. And herein lies the problem: Since you can't check the work, build-and-cover ignores inspections and usually is completed before anybody in the department of buildings gets wind of your renovation. Build-and-cover is a dangerous way to do renovation simply because you will not have any idea of whether the work was done properly.

Many homeowners don't obtain permits or don't obtain all the permits they need. This is sort of like parking for five minutes in a no-parking zone to pick up a package from a store. You break the law, albeit a minor one, and run the risk of a fine.

If you are going ahead without a permit—always risky, but sometimes a worthwhile strategic move—watch to make sure your general contractor doesn't take advantage of the situation and use the build-and-cover method (unless you are absolutely knowledgeable about construction and an ace supervisor).

The Inspector: Angel or Devil?

It's important to become good friends with the inspector who gives you the final okay on the work you're having done. Only the inspector—not the permit office and not your contractor—can reveal hidden secrets of the permit process that may save you thousands of dollars. Permits are the driving licenses of home remodeling, but unlike driving licenses, *when* you apply for a permit is crucial. Applying too early or too late can cost you thousands of dollars.

Take, for instance, this story of a Washington, D.C., couple. They who were told by their contractor, who was told by the city inspector, that because they wanted to replace the lead water pipe that went into their house from the city's outside water meter, they were also required to replace the line that ran from the water main in the middle of their street to their house's meter. To do this, they had to tear up their front porch, lawn, and street, not to mention put everything back together again. Apparently, the couple was told, the D.C. plumbing code required a one-inch pipe from the main to the house; there was only a three-quarter-inch pipe in place at the time. Worse still, the contractor said the couple would have to pay for ripping up the street and repaving the street to the highest possible grade. In other words, the homeowners would have to spend an extra $7,100 just because they wanted to remove the lead from their system.

As you can imagine, this couple wasn't too happy about learning that it was going to cost them an additional seven grand. So they called their contractor back, who shrugged and said, "There's nothing I can do." The couple called their city councilman's office and spoke with a staff assistant who was incredulous but offered no advice. The couple toyed with the notion of dropping a couple of fifty-dollar bills on the floor the next time the inspector came around but dismissed that idea. Finally, they called the head of the Washington building department, who simply sent the inspector back around, as it turned out, the best thing that could have happened.

First, at the couple's insistence, the inspector pointed out the code which said very clearly in plain English that service pipes from the city's water main to a house must be at least one inch in diameter. (The larger the pipe, the greater the water pressure; the city wants to make sure that all residents have good water pressure.) Okay, they couldn't argue with the code.

But the inspector also told the couple that the reason that were required to replace the street pipe had nothing to do with the lead pipe per se. Actually, the inspector continued, the city does want people to replace lead pipes. It's just that code has to be maintained at all times. The inspector pointed out that once they replaced a portion of the pipe outside their house the code required them to replace the pipe all the way to the main in the street.

What does all of this have to do with permits?

The inspector divulged to the couple—off the record—that they could consider not replacing the lead pipe just then (the city doesn't require people to replace lead pipes). Instead, the pipe "might" spring a leak a year later or so, and under those circumstances the couple could then get a *repair permit* to replace the pipe. A repair permit would be different from a construction permit in terms of what kind of inspection the city required. And the city certainly cannot deny anyone a permit to fix a leaking pipe.

This was a valuable piece of information: Some renovation work is best done as a repair, especially if it's an emergency. Emergency work doesn't need a permit in most cities.

Permits are important. More often than not, they protect the homeowner against shabby, incomplete, or even dangerous work. They ensure that all builders bring their work up to code.

But the permit process is inflexible. Once you decide to go by the book, there's no turning back: You must conform to the letter of the law. Even if your imagination wants something that's safe and reasonable—say a decorative window—if the code says something else, forget it.

Sometimes it may be worthwhile not to bother with permits. You run a risk in not obtaining a permit, but life is full of risks that have to be weighed against benefits. One family wanted to replace the screens on their second-floor sleeping porch with casement windows so they could use the porch all year long. Sounds reasonable, right? When the family went to pick up a permit, they discovered that the porch was considered an exterior structure, even though only a cat burglar would be able to enter the porch from the outside. Hello, Kafka, and welcome to our town. What this meant was that the couple had not only to obtain a permit, but first had to get a zoning variance, a prerequisite to enclosing certain exterior structures. Unfortunately, the waiting time for zoning variances was several months. So the couple talked with their neighbors, who had no objection to the work, then said to themselves, "Forget the variance," and went ahead with the work without a permit.

Get Permits Early

As with most aspects of home renovation, it pays to get permits early. In most instances, the contractor will obtain permits for you because he's more familiar with the process. Some G.C.s don't like to get all the appropriate permits until the last minute. So your demolition could be completed, but while you are waiting for the permit, your house looks like Dresden during World War II. If your local permit office is backlogged, the process can take longer than expected. While your G.C. may be the one who goes to get the permits, it's up to you to inspire him to do so.

Sometimes general contractors like prolonged projects. If the contractor is working on several projects simultaneously, he deliberately may not get your permits until "later"; that way, the delay doesn't appear to be his fault. Everybody likes to blame local bureaucracies, and contractors know that you are more likely to point a finger at the city government for this particular problem than at him. If you think your contractor is deliberately delaying, invite yourself to accompany him to the permit office—tomorrow morning.

6 | The General Contractor

(How'd He Get Promoted to General, Anyway?)

WHILE YOU CAN HAVE A SUCCESSFUL renovation without an architect, it's far more difficult to renovate without a general contractor or builder (the terms are synonymous for our purposes), or without acting yourself as the G.C. General contractors do an amazing array of things including

Hiring subcontractors (electricians, plumbers, masons, etc.);

Telling the subcontractors what to do, and sometimes how to do it;

Modifying the architect's, owner's, or builder's plans as unexpected structural problems appear;

Dealing with worker problems;

Buying supplies and materials in a timely manner;

Doing assorted tasks as they're discovered while subcontractors are on vacation, or if subs are lazy or incompetent;

Offering his years of problem-solving experience to your renovation; and

Being a target for your anger.

Architects and builders sometimes wage a hazy war. Builders don't think you need an architect to redesign an apartment, and besides, architects don't know the first thing about mechanical

93

and electrical systems, not to mention structural problems. Architects think builders are sloppy, unimaginative, unaesthetic, and crude in their plans. One Washington, D.C., architect, says, with only faint hyperbole, "A contractor would never consider the neighborhood, the site, the existing building, the building's style, and the client's space requirements, philosophy and aesthetic considerations."

Architects and builders went to different schools. Don't confuse the two.

Anyone, including you, can be a G.C. And there can be advantages to being your own contractor, among them having very close supervision of the work and saving money. But sometimes when you think you're saving money by being your own general contractor, you aren't. Because general contractors have long-term relationships with their subcontractors, a subcontractor is apt to charge a homeowner more than he would charge a contractor. A contractor is probably better at scheduling the job than you are, which means that he may be able to get the work done two, three, or four times faster than you. He knows, for example, that the plumber has to finish roughing-in before the tiler can get to work, that the electrician has to put in the wires (these days, they include electrical, phone, cable, stereo, and computer) before the drywaller can do his job. Coordination is crucial; a misorganized schedule can cost thousands of needless dollars.

Later on I'll have more to say about being your own general contractor.

What Makes a Good Builder?

What should you look for in a builder? Without getting too repetitious, you should use the same criteria you use for architects and design-build firms. Get references and recommendations. Talk to people. A lot. Remember, this guy is going to physically alter the place where you live, which probably is also your most valuable possession. He'd better be capable of doing it the way you want and doing it right. Roughly 75 percent of all home renovators find their contractors through recommendations from friends. That's probably the best way. Another 7 percent rely on the Yellow Pages, probably the worst way. Take a look at the builder's work. Does he excel in basic pasteup work? This is fine if all you want is to turn an unfinished basement into usable space. If you are interested in a sensitive (fancy, stylish, elaborate, refined—pick your synonym) renovation, then only hire a builder who has done that kind of work before.

References are so important that I can't emphasize them enough. The level of checking should be sophisticated, detailed, and current. You should be as thorough as an FBI security-clearance investigation. No kidding. Here's why: Chicago resident Julie Johnson and her husband selected their contractor from a list of "preferred contractors" their bank provided. No interviews with previous clients. "Our bedroom now has a three-inch slant. You can see the horizon over the chair," laments Julie. "The contractor was impossible to get hold of. About three months after the work ended, there was a rainstorm and it started spitting water through the windows. He didn't caulk the windows."

Even if a builder has forty years of near-perfect experience, and even if the builder's previous clients laud the builder, things may be different for you. That's what Penny and Don Moser found when they started to build an addition to their Washington, D.C., home. The Mosers' house borders on Rock Creek Park, a national park in the heart of the city. Their plan was to build a simple, almost boxlike two-story addition. They wanted to add two rooms and create a deck on top of the addition where they could place a Jacuzzi—nothing fancy and nothing complicated. The Mosers selected a builder with forty years of experience and with nary a complaint against him. The man who ran their company was one of those old-world types of general contractors—honest, skilled, a solid citizen.

So why did a twelve-week project take eighteen months? Why was there one disaster after another?

"There were no complaints filed against him," Don says. "No previous problems," Penny adds. The company had letters of reference and bank references—all excellent. Don continues, "Our mistake was that none of these letters was absolutely current. This is one real mistake that we made. We did not say, 'Who are you working for right now? Give me some names and phone numbers.' " Had they gotten that information, the Mosers would have discovered that the company was in deep financial trouble. Other clients were suffering. In fact, as Penny and Don's problems grew worse, the subcontractors gave them the numbers of other clients with whom the Mosers could commiserate.

"That would have made a difference," Don says. "They were an old, reputable company and had fallen on hard times around the time that they started our job. The reasons really never became clear. One of the partners in the company left—maybe he cleaned out the till as he walked out the door, or maybe he was the person who had things under control." He apparently left with a secretary too. "They were in trouble by the time we got involved with them," Don continues, "and hadn't been in trouble for more than a couple of months. The letters of reference that we looked at that were about a year old were OK, but they didn't represent the situation. If you're going to get references, get current ones."

Penny said, "This should have been an omen. The very first thing they did before excavating was take off the cement slab. They jackhammered the slabs, but they never closed the backdoor to the house. I came home and the whole house looked like there had been a flour fight. I asked about that and they said, 'Oh, yes, I guess we should have closed the door.' That was the first time I went crazy."

Before starting the addition, the builder had to excavate the backyard, which he accomplished with a backhoe in two days. "They created a huge mud pit," Don says. "They backed over our neighbor's yard. And then they disappeared. Then the backhoe sat for forty-five days. It was a rented backhoe."

Penny and Don found out later that the company was in the process of going bankrupt and didn't return the backhoe because they couldn't afford to pay for two days' rental. "You'd look at the back, and birds would be sitting on the backhoe. For forty-five days we had this backhoe," Penny says. "Then they tipped over the backhoe on its side."

"By then we knew that something was wrong, but they already had a hunk of our money," Don says. The Mosers couldn't just say good-bye to the company or they would have been out about $12,000.

"Then we started screaming and yelling," Penny says. "I

started going down to their headquarters, but nobody would ever be there. The office would be open, there'd be some guy sitting there, but no one ever knew anything. Suddenly I realized that maybe we had trouble, but then I thought, Well, maybe that doesn't mean anything."

The backhoe was finally repossessed by the rental company. "They were really peeved," Penny says. It was the subcontractor who didn't bother to return the backhoe to its owner because "the subcontractor wanted to get back at the contractor because he wasn't getting paid."

"Consistently, through the entire business, the workmen weren't getting paid," Don recalls. "A new group of workmen would turn up and they would work for two or three days, then they would disappear because they weren't getting paid by the contractor. After screaming and yelling, occasionally some guys would turn up and do something. The hole got dug out by hand in the end."

"They filled a drain with cement," Penny continues. "They weren't paying any attention to what they were doing." Penny had to watch over the workers herself; the contractor didn't supervise at all (he was doing everything he could to avoid the Mosers). "I couldn't ever leave," she says. "Nobody would do anything if you weren't standing there screaming. It was like having kindergarten kids come to work."

The workers the builder hired were the worst money could buy. "They dug the hole but didn't provide any drainage," Penny says. "So, as you can imagine, when it rained the hole would fill up with water. There was a small lake back there. By then the contractor had no money, so he was relying on any two guys with a station wagon to do the work."

The subcontractors, working without much coordination, finally built the wall that was to frame the Mosers' addition. One rainy afternoon, Penny looked out her kitchen window and noticed that "the wall seemed to be getting shorter. The kid had mixed the mortar incorrectly; it was all sand. When they put the blocks in and it rained hard the mortar just washed out."

Penny called the city inspector, who gave the wall a big F. The contractor had to arrange for the whole wall to be taken down and rebuilt. "I took a crowbar and said, 'There's no fixing this, guys. This is the foundation on which the house is built, so we will now start all over. All over.' The walls went up again, and this time they went up with great care because I spent four months of my life sitting with my legs hanging out what was then the door, watching people."

It looked like clear sailing for the rest of the project, despite the company's financial problems. "But when the new carpenter arrived, they gave him the job of cutting the old pipe, but the carpenter didn't know how to do it. He took a jackhammer and broke a major pipe under the house," Penny says. Penny and Don had to find people whom they describe as ex-cons, to tunnel under the house and repair the pipe. "Every day was the same."

"About six weeks later," Penny says, "These guys showed up. They were real smart and they knew what they were doing. I felt really bad because I knew that they weren't going to get paid on Friday. But we never got to find out whether they were paid on Friday because two of them got arrested on the way to our house."

The Mosers finally fired their contractor and completed the job working as their own G.C.

How to Tell If Your Builder Is Reliable

One Washington, D.C., homeowner had just finished his renovation ahead of schedule. He was happily enjoying his new kitchen when, suddenly, his garbage disposal fell out of the sink at 4:00 P.M. on Passover. The plumber was unavailable, and the homeowner was expecting twenty-five guests to arrive in two hours. The homeowner called the builder, who came over and reinstalled the disposal correctly. This is the test: Will the contractor himself come over to fix a problem if he can't find the appropriate tradesman? Ask people who offer references; if the answer is yes, hire the builder.

It pays to see whether the contractor is a member of local or national trade organizations such as the National Association of Home Builders/National Remodelers Council or the National

Association of the Remodeling Industry.* Membership is no guarantee of quality, but it's one of several factors weighing in favor of the contractor.

Break the above rules, and you will regret the mistake as long as you live in the house.

After you're satisfied with the contractor's references and samples, find out how long he's been in business. If it's less than a year, be extraordinarily careful: The attrition rate for general contractors is higher than the failure rate for new restaurants. Architect Ted Fleming paints this painful portrait: "The same story gets repeated over and over again. A talented carpenter decides to go out on his own. He spends nine or ten hours on the job, then goes home and tries to do the bookkeeping. It's almost inevitable that he will fail."

Look at the kind of operation the builder has. Visit his office. Does he have an office staff? I'd personally feel more confident if there were at least one other person on board to handle the paperwork; the amount of paperwork generated during renovations is staggering. The builder's office operation is a reflection of the quality of his field work.

Checking References

Check references. Visit sites that the contractor has worked on before. Don't rely on pictures the contractor shows you. After all, anybody can take pictures of any house they see along the road.

Always hire a licensed contractor. Some states have funds that reimburse homeowners for disasters caused by contractor malpractice, including reimbursements for deposits stolen by contractors who just disappear and for bad work; the reimbursement money usually comes from the contractors' licensing fees. Virginia, for example, has the Virginia Contractors Transaction Recovery Fund, established in 1980. Maryland has the Maryland

*For information, contact National Association of Home Builders/National Remodelers Council, 15th and M Streets, N.W., Washington, D.C. 20005, or National Association of the Remodeling Industry, 1901 N. Morre Street, Suite 808, Arlington, Virginia 22209.

Home Improvement Guarantee Fund. Other states require licensed contractors to post a bond as a guarantee against catastrophe. Only by using a licensed contractor do you avail yourself of these protections. A number of states require licenses from contractors who accept more than a certain number of dollars from homeowners, but you should verify the contractor's license. You'd never consider visiting an unlicensed doctor or even an unlicensed auto mechanic. Is your house any less important than your car? If the general contractor you've set your heart on doesn't have a license—most do not—tell him that this is a good time to get one.

Now, I know that a large number of readers will violate this rule and use an unlicensed G.C. That's not an awful decision if you are working on a time-and-materials basis, paying as you go along and giving no upfront deposit. You won't lose too much money should the contractor vanish (though you still won't have those special legal protections or the assurance that the contractor is legitimate). But for contracts that require a large deposit, an unlicensed contractor can easily become the biggest thief you've ever met.

Call the Better Business Bureau. Contact the Department of Consumer Affairs. Listen to whether the company asks you the right questions. Don't simply go for the highest or lowest bid.

Renovation is such an unpredictable business that following all the rules to find an apparently perfect contractor can still yield atomic-bomb-size messes.

No matter how great the credentials of the contractor or company, construction firms can and will screw up (sorry, there's no other way to put it). Your house may be the one that's in the wrong place at the wrong time. Personnel changes, unforeseen weather, a worker going through a divorce, machinery that chooses now to fail, a strike at the plant that manufactures the fixtures you want—the unexpected happens. If things go badly despite how scrupulously you investigated your contractor, it probably would have been worse had you not been so diligent.

There's some cliché that says, Things are relative. The true meaning of this phrase may be undecipherable, but its application isn't. Let me explain. A wise approach to selecting a general contractor is to interview three. In all likelihood you will have one high bid, one low bid, and one bid in the middle. One contractor will probably appear less qualified than the other two. This is the way the process usually works. Of the two remaining contractors who appear satisfactory in terms of how you perceive the quality of their work (just a perception I should add), most people pick the low bid if it's not outrageously low. But notice something about this process: You quickly and almost automatically eliminate one contractor from the bidding, which leaves you with only two to choose between. Your contractor is then selected from the two. You haven't given yourself the opportunity to compare three contractors because one didn't meet your standards from the outset. The only fair way to choose among contractors is to solicit bids from at least three contractors of equal merit. You don't want to force yourself into a position of choosing a contractor because he was least awful.

If You're Hiring a Firm, Not an Individual

If you're hiring a firm with several general contractors, interview the people who will actually work on the job, not the sales representative, who always appears wearing a coat and tie. (Can you imagine someone working on your house wearing a coat and tie?) Sales reps are smooth talkers, generally honest, but many don't know the right end of a nail. Once you've signed on the dotted line, you may never see the sales representative again. But you will see the foreman or general contractor. These are the people you want to meet prior to signing a contract. Get to know them before they get to know your house. It could be that you don't like some little thing about a general contractor, such as his smoking on the job. Or there may be something about a G.C. that you really like, such as when he sees a pipe being put in off-center, he personally assists the plumber in centering it.

If you can, watch the contractor (or foreman) at work. Spend half a day with him. While you are there, say, "Tell me about five problems you have encountered and how you solved them." That's one of the best questions you can ask. When you've found a G.C. you like, get a commitment in writing that this individual will be on the job all the time. Substitutions are not acceptable.

Make sure that the G.C.—and the firm the contractor works

for—has done this kind of work previously. If they've built great houses, created magnificent additions, installed superb bathrooms, and made wonderful decks but never excavated, don't have them try out excavating on your house. Experience counts. One kind of experience does not translate into another.

It doesn't hurt to ask to look at the contractor's bank references and to obtain company references. Most companies should be open about giving you a bank reference and especially about letting you contact some suppliers so you can check that they pay their bills on time. Companies that won't let you see this information, who beg corporate privacy, probably have shaky finances that they want to hide.

Finally, check the State Licensing Board to determine if the contracting firm is bona fide and that all its credentials are in order.

One kind of company you should be aware of—a sign that's not a priori bad or good—is the firm that specializes in building new houses as opposed to building additions. Technically, building an addition is easier than building a house, although there are some skills needed for building an addition that aren't used in building new houses. (A new house has all its systems integrated from the start, while an addition has to connect new systems to old ones.) When a company takes on a project outside its normal scope of work—new home companies that build additions on the side—that may signal a cash crisis at the company. It can also mean that the company has excess capacity at the moment and doesn't want to leave workers idle, or it can mean that the company is expanding. It's hard to know without asking or investigating. Be cautious when a company takes you on as a client for work it doesn't ordinarily do, even if that work seems easier.

Friends, Generals, and Lovers

Here's a subversive thought: You don't want to be too friendly with your contractor. Sure, it's okay to send out for pizza now and then and have a pot of coffee available for your contractor, but don't carry the relationship too far. Be buddies with your contractor, share jokes, but be prepared. Working with a contractor is like walking through a potentially dangerous neighborhood at night—probably nothing will happen, but then again. . . . Statisticians say the national divorce rate is about 40 percent. Well, if marriages—the most intimate and secure of relationships—can sour, you're not taking any risk by assuming that your relationship with your contractor will have problems. And maybe big problems. If the predicaments become grand enough, this friendship between you and your contractor can turn from joyful to cool to harsh to nasty.

There are exceptions to the rule of shunning friendship with your contractor, and I have to say that the experience my wife, Peggy, and I had with our builder, Paul Locher, has led to a lasting friendship. But I've heard many, many stories in which expectations of a good friendship disintegrated in the blink of an eye.

Enough pontificating from me. Let me tell you what happened to Kathy and Sandy Shapleigh, their basement, and their contractor, Jeff, and then maybe offer some thoughts on whether a bad thing could have been prevented.

From the moment these homeowners and their contractor met and talked about transforming a basement into a recreation room, it was like from the start. "I have this wonderful contractor. I never want to see him again, but I like him," Kathy Shapleigh told me, hinting at something ominous. "He had a laid-back attitude. If he got to the kitchen before I did, he would make coffee." Jeff was one of the most pleasant contractors Kathy and Sandy had ever met, not to mention one of the nicest people overall. "The construction went on . . . and on. It never seemed to be finished, but that was okay. What finally happened," Kathy says, "is that one day we declared victory; the work was "done," even though it wasn't finished."

One requirement of the Shapleighs' project was to heat the basement. "Jeff got this plumber to install wallboard heating radiators in the basement," Kathy says. "After he had put the units in at great trouble and expense, and after I had put in top-of-the-line wall-to-wall carpeting, the plumber said, 'You got steam heat!' "

The plumber had installed a system for hot-water heat. "The only problem with it was that the entire rest of the house was steam heat. The two aren't compatible. But the plumber did what he had to do: Because the location in the basement where the radiator was going was below the boiler, steam heat was not possible," Kathy laments. (If you're interested in why this is so, consult one of those heavy engineering/plumbing books.)

Jeff, the contractor, came to the rescue. Sort of. He told Kathy, "Ralph, the plumber, wants to make things right. He's going to change all your radiators [throughout the house] to hot water for only $1,000." Kathy just couldn't wait to spend another $1,000! But just like the late-night television commercials that sell combination knives and hair dryers, there was more: Hot-water heat is less energy efficient than steam, so Kathy would be paying more for the rest of her life.

Kathy continued. "I was not as happy as I might be, but I wrote another check for $1,000.

"I figured that we were finished," she continues. Hurrah. "We went away to Williamsburg in November for a weekend." And returned. "I was in the kitchen Sunday night and my two kids, ages six and eight, were in the basement. All of a sudden, my six-year-old yells, 'Mom, Mom, there's steam coming from the boiler.' I ran down and scooped them up. By the time the three fire trucks showed up, there was steam pouring from the basement. There was a foot of scalding water in the basement. Nearly boiling water. All of the scalding water from the boiler poured into the basement. The pressure valve had released."

Kathy diagnosed the problem. "The gauge the plumber had left on the furnace had been the original gauge for steam, so the gauge didn't react as the water got hotter and hotter. There was no cut-off." They never put on the proper gauge for hot water heat!

"It was November, and we now had no heat for days," Kathy says.

"A carpet cleaning company came and sucked all the water up. Another plumber came and put another gauge in. The contractor paid for both. When I spoke with the contractor (he's now called 'the contractor,' not Jeff) after nearly seeing my two daughters scalded, he judged from my tone that I wasn't friendly. I was really upset.

"Jeff and I had became friends. I couldn't be nasty to him. Don't be friends with your contractor, because sooner or later there is going to be some unpleasantness, and you don't want to argue with a friend," Kathy concludes.

Was this plumbing disaster inevitable? In hindsight, no disaster is unavoidable, but hindsight is one of those items that can be very, very expensive to purchase.

So here's some less expensive—and equally valuable—hindsight. Develop a friendship with your contractor. I know, I've just contradicted myself. It's helpful to have rapport with your contractor, because after the renovation you might create a real, lasting friendship. But don't let this friendliness cloud your ability to question every aspect of your renovation. You can be both friendly with your builder and critical of what he's doing. Friendship never implies total passivity in any relationship. You have to be willing to let whatever friendship you create suffer for the sake of your house. You have to be willing to trade a new friendship for a house or apartment that you're going to live in for a long, long time.

Kathy said later that had she known that steam heat could not be put in her basement, she would have opted for electric baseboard heat. But she didn't know because neither the plumber nor the contractor told her. And she didn't even know to ask because unless you're an expert, like a contractor or plumber, you simply don't know what the alternatives are at each stage.

Craig Stoltz, an editor of a major metropolitan magazine who survived a kitchen renovation by a contractor who was "generally half-assed about the work," puts the problem this way: "You get into a relationship with contractors. It's not just a contract. My wife and I call it 'identifying with the enemy.' You have to distance yourself to enforce the contract as it's written or you get something you don't want." Craig warns that when the work looks like it's getting too difficult or more expensive than planned, "It's like any human relationship. You begin to feel sorry." But what you have to do, he says, is "enforce the letter of the contract and be vigilant about it. These people are *contractors* for a reason. They think of themselves in terms of the contract they write."

The best way to keep a protective distance between you and your contractor—for the sake of your house—is to ask the contractor what's happening at each stage of the project. You aren't necessarily questioning the contractor's work but appraising the alternatives and trying to understand what's going on. Contractors respect questions (at least the good ones do).

Questions prevent disasters and preventing disasters helps cement good relations with builders.

What you can and must do is ask at each step, "What are the other options we have here?" Ask, "Could we do this differently?" Ask, "Is there another appliance/fixture/system/kind of wiring/type of wood/filter/insulating material that we could use instead?" Make your contractor explain his reasoning about any work that is expensive (such as wood floors), crucial (such as duct work), or necessary (such as heaters). Why did he choose that particular apparatus? Is it compatible with what's in the house now? What can go wrong with the thingamajig he's selected? Is there a cheaper *or* more expensive alternative? Don't just accept, "Yes, but this one is best for the existing duct network." Ask for the alternatives anyway.

When Peggy and I were considering installing a whirlpool tub, our general contractor suggested custom building one for the space. He said it would be no more expensive than an off-the-shelf model. We said okay, but when he said it would be made out of poured concrete, we began to wonder whether it might be too heavy for the floor. A tub falling through to the kitchen wasn't a prospect we relished (we'd seen a ceiling fall in in Peggy's brother's house). The contractor calculated the floor's strength and concluded that it would support the proposed tub, plus water, plus two adults (assuming we didn't get on a pizza kick). We decided to get a lighter, acrylic Jacuzzi tub anyway, to err even further on the safe side. Only by asking, did we even have the possibility of making an alternative choice.

Take a microscope to your contractor's proposals. Like a world-class chess player, examine each option that makes sense. Be firm with your general contractor from the start. Firmness isn't a sign of coldness or antagonism. After all, spouses, parents, children, and lovers are all insistent with each other all the time.

Some Generals Are Better Than Others

Humans make mistakes. If there's any truth to that aphorism, then contractors are the most human of us all. Versatile as they may be, the consummate jacks-of-all-trades, they have weaknesses. Some will be better at things electric; others are woodworking pros; still others are at their prime when it comes to making bathrooms shine.

Hold that thought.

Just because your contractor can perform fifty or one hundred more useful functions than you can, don't assume he can do everything. No matter how dazzled you are by your multitalented contractor, recognize his limits and when you find them, stop him there.

"I have a row house on Capitol Hill [in Washington, D.C.] and my budget was limited, so I shopped around for a contractor. I found this contractor, George Kyros, through a friend. He was going to replace a countertop in the kitchen, put in new bathroom fixtures, make a closet, and paint the outside of the house." This is how Suzanne Pieron began her story of how to discover a contractor's weaknesses.

"He did the kitchen sink," Suzanne recalls. "I didn't look too closely at first, but I later found that he had covered his mistakes with caulk. In the kitchen he nicked a lot of edges and caulked the nicks. When he cut the countertop, one of the corners went an inch too far, so there was a gap between the wall and the sink. Rather than starting over, he caulked. This guy had boxes of caulking and used it to cover any mistake he made. Beware of contractors with fifty tubes of caulk in the back of the truck."

When your contractor isn't doing something the way you want, it's either because he doesn't know how, isn't skilled at that task, is lazy, or wants to save money. I'm not sure which reason is the worst, but whatever the reason, if the contractor won't do it your way, fire him.

When Suzanne Pieron noticed that the same contractor was painting the iron gate of her house black instead of navy blue, as she had ordered, she said to him, "Stop, paint it blue." "Well," the contractor argued, "I've already begun to paint it black, and besides, the other houses in the neighborhood are black and you wouldn't want your house to stand out." Despite the fact that she's a strong-willed individual, and despite the fact this was exactly what Suzanne wanted—something a little different from the rest of the bunch—she relented: "You feel a little dumb that you're paying someone a fortune and he isn't doing it the way you want, but you just want it done." A common phenomenon in the midst of a renovation is that when a contractor starts to do something not according to plan, many contractors are able to convince their clients to continue the work the way the contractor wants. More on this later.

Other painting problems developed. "I had already noticed that only the front side of the cast-iron fence was painted," Suzanne says. "The back side wasn't painted. He said, 'I'll touch that up later.' " As he continued to paint the rest of the outside of the house, "he started scraping a little of the lower part of the house's old paint. I thought things were proceeding along okay.

"Then the paint looked rough. I asked him if he had scraped the old paint off first before applying the new coat. He said yes. But it looked really rough, and I made him do it again. The next day I came home and two-thirds of the house was painted. It still looked rough. He was just painting over parts of the house. By this time my entire front garden was totally destroyed. To him my Korean mint plant was a weed, and he tore it up by the roots and threw it in the street."

Suzanne kept vocalizing to her contractor that the painting didn't look right. "Finally," she says, "he said, 'I'm not really into scraping.' That's when I fired him. He had told me he was an experienced painter. When I hired someone else to finish the job, that person noticed that the cans of paint that George had left around were interior paint."

Suzanne's lesson? "Unless you really know what you are doing, hire a contractor who is expensive and good. You get what you pay for."

Your contractor's view of the world is undoubtedly different from yours. The shared experience of contractors is different than the shared experiences of homeowners who are lawyers, school teachers, accountants, etc. Your contractor may live in a nicer house or a more run-down house than you; you don't know. Your contractor's life-style, fiscal upbringing, physical health, and even personality make a difference in how he approaches your renovation. Just like with regular people. Every office has its assortments of nerds, straight arrows, liars, drug users, goof-offs, sleazes, geniuses, chronically sick. From day to day a person's attitude can change from happy to sour—you may never know what's going on at home.

Craig Stoltz and his wife, Pam, had a simple project in mind: They wanted kitchen cabinets installed. Craig and Pam didn't discuss all the details with their contractor, just the major points like which cabinets they wanted and where in the kitchen to put them. The contractor, according to Craig, "mounted the kitchen cabinets four inches too high, but when we pointed this out to him, he said, 'No, no, they're the right height.' Neither my wife nor I could reach the cabinets because they were so high." How did this problem evolve? Says Craig, "The contractor was six feet four inches and I'm five feet eight inches. (The contractor did move the cabinets.)

What homeowner takes into account the height of his contractor? Certainly almost no one but from now on you should. Height, weight, personality, physical condition—each of the components that make up the contractor affect how he approaches your house. The same is true for the subcontractors the G.C. hires.

Some Generals Should Be Demoted Back To Private

General contractors aren't perfect builders. Neither are they perfectly scrupulous. The majority are up-front, honest individuals. A handful are liars and cheats. Many fall in between these two extremes. Because contractors make a living at this work, they won't plow every penny you pay them into your renovation. Some of the money has to go toward their mortgages, their entertainment, their businesses' overhead, and bribes. From time to time they will cut corners in order to pay for these things—alas, at your expense. It's one thing to keep overhead low, it's another to cut corners. You need to be alert for the telltale signs of cutting corners.

One such sign is stretching out the length of the project. The longer your project takes, the greater the opportunity the contractor has to fund it with other projects he lines up. The money you are paying him may go toward buying materials for another renovation; it may become necessary for the builder to acquire even more clients to keep your renovation alive.

Another danger sign is when only the builder's staff does the work, one person does several jobs, or the builder himself does the work that a plumber, carpenter, or mason might do. The contractor's staff—which includes himself and anybody else who's on the job that day—is usually cheaper than hiring subcontractors. Keeping labor entirely in house is dangerous for several reasons: Your renovation project may take longer (despite the contractor's insistence that he's using the drywaller to do some carpentry because the carpenter won't be available until next week, and this makes everything go faster for you). Your project may become more expensive if you're working on a time-and-materials basis because in-house labor wasn't competitively bid. But the gravest danger is that the quality of work might suffer when the builder's staff does the work that should be reserved for a specialist. When someone has built four hundred doorframes, you can be sure he'll build yours correctly; when someone has *helped* build only three doorframes, you may become the owner of a crooked doorframe. Specialization is in your interest. Don't let the builder cheat you out of it. Indeed, some specialized subcontractors are required by law, usually mechanical, electrical, and plumbing work. If you see your builder performing these functions himself, then the mechanical integrity of your house may be in jeopardy. (There's a danger that the G.C. may skimp on materials too.)

General Contractors Shouldn't Be Their Own Armies

Some projects are based on the notion that one individual, the master carpenter, will be both the general contractor and the person doing most of the work. On the face of it, it sounds like a good idea to have a talented craftsman tackle your renovation hands on, leaving only the plumbing, masonry, and electrical work for others. If you've investigated this individual's ability to do the work, you've probably discovered that he takes a keen interest in the houses he works on.

At least that's what Beth Farker thought. In fact, despite a series of unpleasant incidents and depressing weeks, "My kitchen came out beautifully," Beth says. Alas, when you are relying on a single individual for a substantial portion of the work, your life is subject to the vagaries of that person's personality and life. The work will be slower than if a team were involved. How much slower is something you won't find out until you are practically out of your mind. "I consulted people for references and they said he was super," Beth reports about her search. "His references said that he works day and night, and the only problem is that he gets exhausted from working so much." The carpenter started the last week of September, in plenty of time for Beth's kitchen to be finished for her family Christmas dinner. "He said he would be done in a couple of months. In terms of time, the only limits were the countertops because they were going to be made of granite and would not be ready by Christmas. The carpenter said he would put plywood on the countertops as a temporary measure.

"The carpenter started doing a fine job, coming in every day," Beth says. "After about a month his behavior started to get odd. He started to show up irregularly. He said, 'I'll see you tomorrow,' and then did not show up. When I would call him on it, he got upset with me. He always had stories about the other people who were making his job miserable. They didn't pay him, they made unreasonable demands, etc."

One particularly odd event occurred. Beth went away for the weekend, leaving her house with the carpenter.* When she returned, there was a note that said, "Your wood is in the basement." But Beth didn't see any wood in the basement. "I called him," she said. The carpenter asked Beth if she didn't see any firewood in the furnace room? Beth reported, "He said that somebody had come to the door and told him that I had placed an order for firewood. The carpenter paid for the firewood," Beth said.

The carpenter kept giving Beth long stories about his personal difficulties: "He told me how he had been taken by this mover,

*Oddly, people who would never invite their aunts and uncles to stay overnight give workers keys and free rein to their houses.

and asked, 'Could you please pay me the next installment ahead of time?' Then he would repay me by not showing up for a week. He started giving me gifts from the Scratch 'n' Dent. He would bring me a treasure like a kid trying to get mom to love him when he's really a terror. He would repair my broken things without asking. You wanted to beat him over the head but couldn't."

Finally, on Christmas Day, all the carpenter's tools and pieces of wood remained on the floor. There was a hole in the kitchen floor. Just as Beth was ready to kill, a basket of fruit arrived from you-know-who. Christmas dinner was, in Beth's words, "hell on the hostess."

Unfortunately, there's no practical lesson to this tale. Beth loves her kitchen, but her carpenter drove her crazy. Part of the problem was that there was a real deadline—Christmas dinner—looming in the distance. Another part of Beth's problem was that when you hire a multitalented tradesman, you are apt to get a multipersonality tradesman too. When you have a single individual working on your project, there's no one who can substitute for him. You are at his mercy. All this makes living through the renovation emotionally stressful. "The renovation was incredibly inconvenient, of course. But I think the emotional distress was the worst," Beth concludes.

Watch Your General Contractor's Budget

I'd like to dwell a little on one aspect of something I mentioned earlier: contractors who use one project to pay for another. This is a story about a G.C. who was working on a house in Potomac, Maryland. As far as the homeowner was concerned, everything was going swimmingly—the foundation and framing were done correctly, the roof was completed on time, the windows and doors were set in just as specified. But apparently the builder decided that he didn't have enough money to make payments on prior work (another project that was at some unknown stage of completion). The builder decided that he needed his next *draw*. A draw is part of a payment schedule for work performed under contract. Generally, the money paid is broken down into draws: completion of framing—draw, close in drywall—draw; completion of trim—draw. Draws are payments made to the general by the homeowner. The faster the work is done, the faster the G.C. is paid for the latest component. The G.C. in Potomac needed money, so he decided to rush the job so that he could get his drywall draw prior to the rough ins of the mechanical systems. As far as the homeowner was concerned, he thought the drywall was up. Great! So he wrote the check. When the mechanical tradespeople arrived, they had a different point of view. They had to break down the drywall. The mechanical subcontractors, who had been promised a clean, open work area were upset. They didn't break nicely into the drywall, they smashed into it—and charged dearly for this event.

But this wasn't the end of the story.

The builder was incapable of completing the job under budget because he had to redrywall 50 percent of the project. This ruined the builder financially, and the homeowner got his house finished in sixteen months instead of eight. The project took twice as long because the builder needed to line up other projects to pay for this one in the meanwhile.

From your perspective renovation is about living. From the contractor's point of view, it's about making a living. These perspectives can come into conflict.

You have to observe what's going on with the general contractor's past and current projects. Some G.C.s use what could best be described as a pyramid scheme: present cash outlays are financed by future projects. This system isn't limited to small contractors; some large firms do it too. Contractors are sometimes unable to determine what it costs to build a job,* and consequently say to themselves that they will make it up on the next job. The result is a conflict of interest as they try to rush the job that they have lost money on (the current project) in order to begin work on the next project and start drawing cash from it. Once this starts, a G.C. can get deeper and deeper into a hole.

Builders easily get overextended.

Unfortunately for the homeowner, this is very common. One builder, who insisted on anonymity, estimates that as many as one out of every two contractors has used this formula at one time or another.

*The Pentagon is familiar with this.

Keep an eye out for builders who want to rush or slow down a project. Inspect each stage of your renovation as if it were the only project underway. You will help ensure that the builder isn't cheating you through a cyclical pyramid scheme.

Put on your thickest glasses and examine the project as if you were a combination Internal Revenue Service auditor/Chairman of the Joint Chiefs of Staff.

What Makes a Bad Client

Like architects, builders and contractors have their criteria for what makes a bad client. In the words of Mark Richardson of Case Design in Washington, "Communication is the key. Nine out of ten of our problems are because of communication, not because we speced a window wrong but because what we were doing was misunderstood by a client." When the homeowner doesn't know what or why something is being done, this often leads to friction between contractor and homeowner. Often, however, the contractor won't volunteer to explain, so the onus is on the homeowner. Ask, "Why is this being done with this material?" Ask, "What have you experienced using this on other projects that I should be aware of?"

Hand in hand with communication goes "cooperation," says Vincent Menonna, an electrical contractor in New Jersey. You have to be ready to cooperate, he says, because there will be "gray spots in the plans that the architect drew. How do you fill in the gray areas? You want to do this in ways that don't cost you money and time." To accomplish that, the homeowner and builder have to be ready to make accommodations.

Having doubts about your relationship with the contractor and subcontractor is also certain to cause problems. According to Mark Richardson, "If you have a serious doubt about the relationship, resolve it or don't proceed. When you have someone in your home all the time, the relationship is important," he says. You must feel comfortable with what is going on.

On the flip side, a survey taken by the National Association of Homebuilders indicates that the biggest pain for contractors is homeowners who keep changing their plans.

How to Be Your Own General Contractor

Even if you don't elect to oversee your entire renovation, there will be times during the project when you will have to be your own general contractor. You will have to order supplies. You will have to tell workers, "Yes, do this; no, don't do that,' as well as the dread, "That was done wrong; re-do it." Your general contractor may be at the dentist or recovering from a hangover—or both—while a decision has to be made at your house. You will have to coordinate the renovation while you contractor is on vacation or is sick for several days. You will have to inspect work and authorize the next steps. In other words, there will be times when you will be in charge.

Being your own contractor or, I should say, acting contractor is not as urgent a challenge as responding to a distraught flight attendant's query, "Does anyone on board know how to fly an airplane?" but it's close. If you know some things—perhaps you played with the computer game Flight Simulator—you may be able to land the plane safely. If you've never tried the game, you and the passengers have no hope. So it is with your house.

The more you know about being an acting general contractor, the better the chances that you will be able to survive an emergency. (Of course, if you *are* the general contractor then you had better pay extra-close attention to this section.)

Much of this book is indirectly devoted to describing what a general contractor needs to know. Chapter 9, on the people who do the work, talks about each subcontractor's functions and Chapter 7, about planning your renovation are components of this. But the information here is designed to give you a feel for what a general contractor would do, or rather what you should do when you are the acting general contractor.

You're going to have to plug yourself into a different network, a network of tradespeople, good old boys, laborers, and many other people with whom you may not ordinarily associate. If you're in sales, if you're an attorney, a physician, or an editor, you're going to have to abandon your normal work attitudes and adopt a brand-new mindset. You can continue to go sailing on weekends and shopping on Thursday evenings, but if you are going to be a general contractor, you will have to evolve into a different being, at least for a while. To manage workers, you have to know how they work, what their needs are, and how far you can push them.

Figure Out the Job and Find the Person Who Is Best Suited to Perform It

One of the most basic principles of being a general contractor is knowing who is supposed to do what. Sounds simple enough. Electricians do wiring, painters put colors on walls, plumbers put in pipes. Right. But wrong if you think figuring out who is responsible for a particular job is always obvious. Even general contractors get it wrong. Many have to be educated on the job, and you are their teacher. Walking with contractors through the house before, during, and after can make both you and the contractor smarter about what's supposed to happen.

Paul Locher was trying to finish exterior landscaping treatments on a Virginia house before the owner moved in. The last work that needed to be completed was the leadwalk—a sidewalk that goes from the street to the house. The leadwalk required a couple of stairs to get up a slight hill. Paul's crew started digging for the stairs and encountered a rock. Not just any rock, but a rock about the size of a Volkswagen. This wasn't something that could just be shoveled out of the ground, rolled over, and put in the truck. But rather than removing the entire eight feet of rock, all that was needed was to take about two feet off the top of the rock. Paul procured some cold chisels and twenty-four-pound sledgehammers for his crew.* The laborers Paul hired took the cold chisels and sledgehammers and pounded away at the rock. They were taking off little, tiny chips about the size of half dollars and were working as hard as they could. When Paul came back to inspect the work an hour or so later, he saw that they had removed only about four inches from the rock. At this rate it would have taken a week. Paul called over his favorite stonemason, Leroy, who was working on the same job doing some brickwork. Leroy stopped laughing at the workers, walked over to the

---

*These tools cost $65. Renting air compressors and jackhammers would have cost about $700 per day.

rock, took the twenty-four-pound sledgehammer (no chisel), dropped the sledgehammer on the rock, and took off football-size chunks. It took him about six swings to remove eight inches of the rock. Unlike the laborers, who were hitting the rock directly, Leroy knew how to hit the rock at the correct angle. Although you'd think smashing a rock is a job for strong unskilled workers, it wasn't so in this case.

Plumbers, for example, are not pros at laying down bathroom tile, even though you might assume that everything having to do with the bathroom is their province. Painters aren't always good at plastering. Electricians don't necessarily know the best ways to route phone wire through your house (and they're certainly not good at centering outlets.)

How do you determine the best person for the job? In this era of ever increasing specialization, one good rule is to find a contractor with the name of the function you want performed in his title—for example, tile layer, telephone installer, roofer, chimney sweep, and so forth. There are going to be times when specialists don't exist or you can't find one. Who ever heard of a hot tub installer? But when you can't find a specialist, ask a likely candidate this simple question: Have you ever done this before? Greenhorns are not invited to work on home renovations. If this is the first time that the carpenter has installed a mirrored ceiling in a bathroom, find someone who has installed a mirrored ceiling. Let the carpenter you've already hired do everything except put in that ceiling. You may hurt his feelings, but so what; hurt feelings are a lot better than glass all over the floor. Same goes for removing the roof to have your grand piano lowered in. If you've never seen a piano lowered into a house and the person doing the lowering has never seen it, you are taking a larger risk than I even want to think about. A general contractor can supervise a worker who's doing something different for the first time, but only if that general contractor—or you as the general contractor—has done very similar work before.

It goes without saying that you should take care if you hire your owns subcontractors. Investigate them carefully. "We needed a plumber," Julie Johnson of Chicago, Illinois, says. "I called these numbers from the display ads in the Yellow Pages, and they were all the same company. 'Oh yeah. We're all the same company,' the plumber who came over said. The plumber we got was a professional wrestler in Mexico during the summer. He left the work half done, and we didn't want to see him again."

There is a better way to find your own subs. "Our neighbor is an electrician," Julie says. "Everyone he recommends is great and affordable. I wish we'd used recommendations as a way to find tradesmen from the first day."

Uses of the Telephone

When you have a question without an answer, call. When you are substituting for your general contractor while he is sick, for example, and a subcontractor wants to do something that seems out of the ordinary, order him to wait while you get an answer.

Never feel rushed to have a particular job started or completed if you aren't sure about that job. Trust your instincts in this matter. If a plumber wants to go ahead and put a pipe in and you aren't sure whether the pipe should be one-half inch or three-quarter inch, but the plumber says, "Yes, it's supposed to be one-half inch" and it so happens that one-half inch is the gauge of pipe he has with him, tell him to wait. Call the general contractor to find out. Call the manufacturer of the appliance to which this pipe is being installed to find out. Call a hardware store if there is one you trust. Call the distributor. If you are still unsure, tell the plumber to hold off. Which he will unhappily do. If it's the wrong gauge pipe, the general contractor would have told the plumber not to install it anyway, so your request is not unexpected.

When you don't know the answer, say so. The classic example of pretending to know what to do comes from Eric Hodgins's *Mr. Blandings Builds His Dream House,* published in 1946, in which Blandings, a New Yorker, builds a country house. While "overseeing" the construction, a subcontractor asks Mr. Blandings, "On them second-floor lintels between the lally columns, do you want we should rabbit them or not? From the blueprint you can't tell which way they're supposed to be." Figuring that something without is going to be cheaper than something with, Blandings responds, "No, I think that's something we needn't bother with, come to think of it." In the next instant, Blandings is nearly clobbered by the shower of lintels the workers removed to meet his request. (Lintels are horizontal structures that support loads over openings such as windows and doors.)

Whatever happens, resist the pressure from a subcontractor to proceed when you aren't ready.

The Nature of Command

If you give an instruction to a worker that is contrary to the way that worker usually operates, the moment you go away, the worker will go back to his bad habits. Especially if your instruction involves extra effort of any kind. For example, on hot, humid days, laborers are supposed to spread mulch around the landscaping. No one wants to work outdoors. The laborers don't care where the mulch ends up or how thick it is. All the workers want to do is get back inside where it's cool.

The G.C. knows the mulch has to be laid, but he's also responsible for making his workers feel okay. A good manager will split the crews in two and have some people working outdoors while the others stay cool inside, then reverse positions.

Be ready to show your subcontractors how you want something done. This doesn't mean that you have to go up on the roof and show the roofer how to nail. Information as basic as drawing a diagram may suffice. This shows that you've done your homework, and the workers appreciate that.

However, if you want the soap dish in a particular spot in the shower, you must mark the exact location. If you want an electrical outlet here but not there, you'll have to specify that information too. Trust the skills of your subs, but don't trust their ability to read your mind or anticipate your needs.

There will always be questions from workers like, "How should I make a hole in this foundation wall?" You have to be prepared to give an answer. Assuming you know the answer, say, "Let's try first with the hammer and chisel that we have here, and if we have to we can get an air compressor later." While this book is the best on dealing with workers as people, *The Home Remodeler's Combat Manual* is not a substitute for technical books. This book tells you how to deal with the "who" elements of home renovation, not the "what" aspects.

By reading about renovation and talking with other homeowners who've gone through the process, you'll become very well educated about construction. If you can teach a worker something that he hasn't done before, the worker will respect you and do practically anything for you, especially if it makes his work easier or faster.

When overseeing carpenters, the easiest thing to do is make suggestions rather than command. "Have you thought about planing this door to make it fit? Have you thought about removing some drywall here?" Figuring out the geometry to allow a doorframe to be built has a positive impact on carpenters. If there's an electrical wire in the carpenter's way, you as the acting general contractor could say, "Here, wait, I'll move the wire."

The best way to demonstrate your command as a general contractor is to do hands-on work from time to time. Take a hammer and pound a few nails. Mix some cement. Hold the window in place while the carpenter secures it. Measure a space. If you dare, take off the tie and jacket and get dusty! The more you appear to be like a G.C., the more your subcontractors will treat you like one.

Be Prepared for Stupidity

You have to learn how to control the situation, and to teach the workers both how you see the job and why you want something done a special way. This applies to every aspect of renovation. When you're putting recessed lights in the kitchen ceiling, you'll have a better chance of successfully getting the lights where you want them if you explain to the electrician that you read the newspaper at the kitchen table, which is going to go against a particular wall. Thoughtful placement of a window may not seem important to a subcontractor who's trying to modify what the blueprints ask for after finding a pipe in the wall where the window's supposed to go, but if you want to be able to see a mimosa tree from your bed (or block out the neighbors' garage), let the worker know.

Paul Locher was managing a project in which he needed holes

to plant trees in. So he hired a laborer to dig them. The worker seemed congenial, intelligent, and eager to please. And he paid a lot of attention to detail—to a fault. The worker's instructions were to dig holes two feet by two feet and eighteen inches deep for these trees. It took him about three hours to dig each hole.

The problem turned out to be that the worker didn't look at the round balls containing the trees' roots and decided to literally interpret the instructions for the size of the holes. The worker had dug three perfectly squared holes. No wonder it took three hours!

After hydrogen, the second most common element in the universe is stupidity.

Be Prepared for the Unexpected

As a general contractor, you never know what's behind walls until you knock them down. Short of bombarding your house with a massive dose of X rays, there's no way you can anticipate everything that you'll encounter, or even *anything* you'll encounter. Take the case of a homeowner in Maryland who was doing a little excavating in his basement. He tugged on a wire (after having cut the power) to see what this particular wire connected to. What the wire connected to was the tail end of a blacksnake.

A Canadian homeowner found an unsuspected septic tank in his backyard where excavation was supposed to take place. I'll leave it to your imagination to figure out what's involved in removing a septic tank.

It's not okay to admit to yourself that you're not handy.

Julie Johnson, home renovator

7 | Planning Ahead and Keeping to Your Plan

YOU CAN'T HAVE EVERYTHING. With any renovation, there are always tradeoffs, and most of the time these tradeoffs have something to do with money. One Washington, D.C., home renovator, Michael Dolan, put it this way: "You can control some of the elements of a renovation but not all." The three variables that we, as homeowners want to control are:

Price
Quality
Speed

As Michael said, the homeowner gets to pick two of the three; the third is up to the builder. In other words, pump in enough money, and you can have a fantastic job that's done at amazing speed.

But that's going to be a lot of money.

If you're not willing to spend more than you have to, then expect to see less favorable results in either quality or speed.

Enough of complex mathematical relationships. At the outset you have to decide what type of renovation you want and then communicate this to the builder (I'll describe the types of renovations in a moment). If you don't do this, you'll end up with unrealized expectations, poor quality, or a never finished project.

122

Determine what is important to you and what is not.

Deciding among these three variables requires Solomonlike wisdom. All are important. Selecting which element to concede to the contractor is as subjective a decision as choosing a mate. Each case is different. If you have to move soon, obviously speed is crucial, but what about price and quality? If you entertain a lot and need lavish living and dining rooms, quality is the variable that you must meet, but at what cost and how quickly? If you've just learned that your wife is having triplets, you're probably going to have to conserve your funds for diapers, cribs, and other necessities of parenthood, but do you care when the work is completed and how fancy the new rooms are?

No book can tell you how to select from among this mix, but it's important to be aware that you may have to make these choices. One way to approach this dilemma is to not weigh each variable, agonizing over whether you can afford this or that, or whether constructing this or that will take an eternity. Instead, think about what you want out of your renovation. Answer the question, Why am I renovating? Thinking about your needs instead of your house will help determine which two variables are most important.

Timing

There's an old joke that goes like this:

COMEDIAN(to the audience): Ask me why I'm the funniest comedian in St. Louis? Go ahead, ask.
SOMEONE IN THE AUDIENCE: Why are you the fu . . .
COMEDIAN(interrupting): Timing!

Call it timing, planning, or the order of things to be done, the order in which you do things is as important as the work itself. A good contractor will know how to time or pace the work, but because home renovation is such an unpredictable, unevenly tempered animal, even a good contractor won't get the timing 100 percent perfect. Sometimes you simply can't know what should come first. Should you put the hot tub in the bathroom before the bathroom walls are completed, so that the tub fits in the bathroom easily, or should you wait until the bathroom is practically complete? Bringing in the whirlpool tub early guarantees that it will fit through the doors (since the doors aren't even in place), but leaving the hot tub in the bathroom during construction means that it will be in the way and could be damaged. Waiting until later may present a logistical dilemma, that of squeezing a twenty-seven-inch tall tub through a twenty-eight-inch doorframe takes more than a modicum of finesse.

Should you tile the kitchen floor first or put the appliances in first? And what do you do with the refrigerator that arrived two weeks early? Which is better to do first: painting or putting in the ceiling fixture? Should you take advantage of a winter sale and have the HVAC company put in a central air-conditioning system before the rest of the renovation starts?

There isn't always a clear answer to which comes first, but that doesn't mean you shouldn't think about the order in which work is going to be done. You should. You must. In your mind and out loud with your contractor, examine over and over again the order in which work will be done. Otherwise you may find yourself quite literally painted into a corner.

Order Ahead

Order everything you're certain you want as soon as possible. It's inevitable that one or more items are going to be out of stock, unavailable, delayed in transit, stolen, or ruined in shipping. Or, you may hate them when they finally do arrive. It may turn out that you're unexpectedly constrained in your choice of materials. Gail Ross, who was remodeling just one room in her suburban Maryland house, found herself up against the local homeowner's association, which told her that there was only one kind of window she could install. "The biggest problem I had was not with my contractor but with my association, which has all these rules," Gail says. The association made her "get windows that are two to three times less efficient than what I want." So much for the environment. Worse still, "You can only get them from one source, and they were backed up. So the contractor had to wait for the windows to come."

When Michael Dolan remodeled his house he was keenly aware of the ways in which delays can frustrate a project. So he "hired the services, but bought the materials. All the guys had to do was show up with their tools." That's certainly an excellent strategy for avoiding delays (and, as we'll see later, for saving money too). The more you are prepared to visit tile stores, lumberyards, plumbing supply shops, quarries, and hardware stores, the more control you will have over the timing of your project.

Ideally, at the point when you apply for permits you should order all your windows and exterior doors. It will take four to six weeks to get those materials, and about four to five weeks to get this phase of the project under roof. You are going to be looking for those windows and doors about five weeks after you start the project. If the door or window is a custom job, you have to allow eight weeks.

Without windows and doors, you won't be able to pass exterior inspection, what's called *close-in.* If you aren't closed-in, you can't get permits for the mechanicals or get inspections for rough-ins of electrical and plumbing work. In other words, if you haven't completed the outside, work can't legally start on the inside (in many parts of the country).

Your builder should be taking care of this. If he isn't, he knows he will be fostering delays. Some contractors think they can get off-the-shelf units at the last minute, and they may. But these off-the-shelf items are lower-grade materials, which cost the contractor less. No single local store will stock every size window or door that you want, so beware the contractor who wants you to change the size of a window to meet the size that's available. It's easier—and more profitable—for a builder to use an easy-to-find window than to meet the specifications of your plan. So he'll reshape the window to fit what he has.

As soon as you can (even before the work starts) decide on

Tiles;

Floor type;

Light fixtures;

Kitchen appliances;

Bathroom fixtures;

Windows;

Doors;

Wood siding;

Brick siding;

Hardwood flooring;

Ceramic tiles;

Kitchen cabinets;

Specialty items like faucets, sinks, tubs (allow eight to ten weeks, if not longer); and

Where you want to take your next vacation.

Plan your construction around the specialty items being delayed. Generally, the better quality the materials or appliances, the longer they will take to get to you. Anticipate what might not be in on time, and plan to renovate around those absences. It's time to be pessimistic: Think, "What if materials x and y don't arrive on time?" Pry from the builder the status of materials to come.

How difficult can it be to locate quality materials? To get the best price for hardwood flooring, one builder had to place an advance order for 100,000 board feet just to get on a list so that he could draw on a supply for one project. There are nightclubs in New York that are easier to get into.

When the materials you want arrive, tell the builder to have the materials held in the warehouse. Not everything that has been ordered has to be delivered at the same time.

Materials are important to have on hand, but so are the people who put them in place. When Paul Locher subcontracts work, he calls a month ahead of time to tell subs to make room in their schedules. Ask your G.C. whether he has people available for the work that's going to be done four weeks from now, not just a couple of days from now. Materials and tradesmen go in tandem. The plumber should be called four weeks ahead of time, then two weeks ahead of time, then two days before he's expected. Each time the G.C. talks with the plumber, the schedule can be modified slightly. The original target date doesn't have to be stuck to (keep this in mind if you are your own general contractor).

Perhaps the only thing you can procrastinate about is paint color, because it usually takes half an hour to order paints.

You can always change your mind about fixtures, appliances,

and kinds of materials—if you have to. But if the materials aren't on hand when you want the work done, the work won't get done.

Unique Tiles and Other Artifacts

Tiles are in. Not those green institutional tiles that remind us of the 1950s, but glorious hand-painted, hand-fired tiles. Tiles designed in Israel, crafted in Portugal, collected from Italian ruins, secretly exported from Iran, patterned in China—these are the kind of tiles that are gracing American homes today. The spectroscopic diversity of tiles and their artistic heritage lets home remodelers bestow personality to their kitchens, bathrooms, living rooms, hallways, and bedrooms. There are unlimited colors and patterns available. Tiles are the most durable component of many homes; you can expect tiles to last 1,000 years or more. Long enough.

The only drawback with decorative tiles is that they take time to acquire. (Ordinary tiles are usually available swiftly.) Unless you live alone, agreeing with your spouse on what tiles to choose or schlepping from tile store to tile store weekend after weekend will take a long time. Do you want the same tiles on the floor and on the walls? On the walls and inside the shower? The same tiles in the kitchen that are in the pantry? Will a tile you like go with the color of the sinks you think you want? Frequently, when you think you've decided on tiles, on the way out of the store your eye catches another type that you like even more. Believe me, deciding on tiles takes a long time, not to mention considerable patience.

But resolving the tile argument is the fast part. There's an inverse relationship between how much you like a tile and how long it will take for the tiles to arrive. If your tiles are arriving from Italy, Portugal, or another European country, be prepared for a strike to delay their arrival. Maybe by years. In other words, be prepared to find new tiles.

Herein lies a moral: Order your tiles early, early, early! Let me tell you a story about tiles. Once upon a time there was a Colorado husband and wife who, after weeks and weeks of arguing, discussing, coaxing, cajoling, and kissing to make up, agreed that some Israeli tiles they found in a catalog would be perfect for their master bathroom. You know, the bathroom with the whirl-pool tub and double sink and alcoved toilet. The tiles had animals, desert scenes, ancient birds—they were art. So the couple marched to the store that carried the tiles (the store was part of an exclusive national chain). The day was a Saturday (wrong day to go tile shopping). When the couple got the attention of a store clerk by waving expensive tiles around in a dangerous fashion, the couple was told that these particular tiles were no longer being manufactured. In other words, it was out of stock, discontinued. Well, not exactly discontinued: The Israeli designer of the tiles had died recently, and the factory in Portugal that made the tiles was on strike, but there was some possibility that the tiles would be made again in about a year. *But*—the store had about twenty tiles in inventory, just enough to do the backsplash around the mirror. And maybe the couple could find more at some of the other stores in the chain. Maybe. What was the couple going to do?! They knew they couldn't agree on another tile for the bathroom. Would this mean that they would wallpaper their bathroom? Wait for a year until the factory maybe made more? Somehow, however, a miracle occurred: As the couple was leaving the store, another couple was coming in to return two boxes of the tiles—the exact tiles. In scenes that the store didn't have! The couple bought them on the spot, but even the two boxes weren't enough. A few weeks went by, and not a peep from the tile store. Calls and calls and the tile store kept saying, "Well, maybe something will turn up this week." This week. This week. So the couple decided to experiment: They called sister stores in California, Philadelphia, New York, and Florida, and guess what? Between all those stores they found enough tiles to complete their bathroom.

Tiles are not the only material that can slow down a project, but they are an important one. The thought to keep in mind throughout your renovation is that if there are any materials that you are responsible for acquiring, track them down early.

Any material you want that's out of the ordinary, is manufactured in a small dictatorship, or has to be custom crafted is going to take a long time to acquire. Plan for those items far—years, if possible—in advance. Acquiring special items for renovation takes longer than having a wedding dress made.

Materials

It's useful to know which materials are supposed to be used by particular workers. There's considerable latitude when choosing materials. If you don't specify what you want either in the contract or just before the specific work starts, the G.C. or tradesman will select for you—probably a cheaper material than you would like.

MASONRY

Masonry can be cinder block, brick, stone, or slate. Each of these looks different and performs differently. If you're using cinder block, it doesn't matter what the mason picks, as long as the dimensions are correct. However, if you use brick, there are hundreds of different kinds of brick available. Without a word from you, the mason will likely chose the cheapest one, or one that has no character, or one that he likes, or the one that's available. If brickwork is going to be visible at all, you probably want to learn about bricks and select the type you like best. You wouldn't leave the choice of paint color to your painter, would you?

When it comes to stone, generally local fieldstone is the least expensive. You don't want to go out and select stone from other places because the farther away you go, the more astronomical the price becomes. Italy comes to mind in this regard. One ton will do a thirty-five-square-foot area. Unless stone is a focal point of your exterior architecture, let the mason pick the stone.

Slate is used in many renovations. Unless you live in Vermont, Georgia, or Pennsylvania, you don't have a local slate. Because most slate is shipped, the various colors cost approximately the same when compared to fieldstone prices. Pick the color you want, unless you want the mason selecting red or green or black or gray for you.

CARPENTRY

Lumber selection is the most important decision a carpenter might make for you. The varieties of wood are almost as diverse as the faces of people. Despite the endless variables, there are some elements you should keep in mind. Fewer knots look better and make the wood stronger. *Joists,* in fact, cannot have any knot more than one-third of the width of the joist.* Unfortunately, grading lumber is about as precise as grading history term papers. It's all done by eye. Paul Locher still has problems with lumber companies because lumber switching is so common: Unless you carefully inspect the wood you order, chances are you won't get what you requested. If all you want is common lumber to stain because it will be exposed, go with clear lumber which has known knots and straight grain. If you want wood with a little character and you ask for B-grade lumber, which specifies only a few knots, none larger than one inch, you find C-grade which has more than a few knots up to one and one-half inches wide, has been delivered. Grade switching puts money in the seller's pocket. Watch out for it. Don't automatically request A-grade lumber because you may pay a high premium. In addition, A-grade isn't always available, so the lumberyard may give you a lesser grade and call it A-grade.

If the wood is going to be painted, the grade makes less of a difference. However, with wood used in large areas such as siding that is going to be stained and which is fairly expensive to install per square foot, the grade does make a difference. In these cases, someone should assure you that the grade is accurate because the price can change by 25 percent for each grade change. You should get what you pay for. On large orders for siding, for example, swindling is rampant. If you spot lesser-grade wood, send it back. (As always, deferring payments makes this possible. Paying on delivery doesn't give you a chance to examine your stock.)

Wood trim makes a house or apartment look complete. The latest material for cheap production is what's called *finger jointing,* scrap lumber joined together like the fingers of two hands and then milled much like a regular piece of wood. These kinds of trim are the fish sticks of wood. If your wood is going to be

---

*Joists are placed on beams and extend at right angles from the beams. Typically, joists are two-by-eight inches or two-by-ten inches, with the longer dimension positioned vertically. Joists are spaced at either twelve-, sixteen-, or twenty-four-inch intervals. A two-by-eight-inch joist should not be longer than 12 feet without a supporting beam; a two-by-ten-inch joist should be a maximum of fourteen feet without a supporting beam. The subfloor is placed on top of the joists.

exposed (stained, not painted), don't get finger-jointed material. Finger-jointed wood comes in regular lengths but is made of small sections; good trim comes in long pieces with no joints. Although it's rare, be on the lookout for finger-jointed structural 2×4 supports—a real danger. Many engineers don't consider 2 × 4s to be structural, anyway.

MISCELLANEOUS MATERIALS

Never let yourself be talked into using "odd" materials. Don't get creative with structural materials. Use cinder blocks and concrete for foundation walls, not plywood. Sand foundations are laid between plywood frames, but who's going to warranty the life of plywood that's underground? Stick with materials that work, even if the alternatives cost less.

Nails

Carpenters and roofers are the tradespeople who most often handle nails. However, practically every worker deals with nails at one time or another. It's during those few crucial minutes when they do that these workers—masons, carpenters, plumbers, electricians—become *nailers*. While nailer isn't a profession, it's such an important job that you ought to think of these tradesmen as nailers too. Nails are the glue that hold houses together, your house included.

A Very Short Lesson on Nails

A worker can do incredible damage by using smaller nails than a job requires. Every job needs a certain size nail. There are dozens of varieties. Memorize the following or refer to these pages when the time comes. Nail size is measured in *pennies*. Each penny is one-fourth inch, designated with a *D*. For framing, you need at least a twelvepenny nail. To do siding, you need at least an eightpenny nail. Check with the carpenter before he begins his work to see what size nail he's using and have him explain why he's using that nail. It's your prerogative to ask that the worker upgrade his nails, period. He will say that you don't need a better nail. You should respond, "Go ahead, use a larger-penny nail." There are always better nails. If the worker is using a 6 D galvanized nail, implore him to use an 8 D *spiral nail*—spiral nails are grooved and stay in material better. Get to know your nails.

The rule of thumb is: A nail should penetrate between 50 and 100 percent of the way into the nailed surface. If you are nailing a two-by-four to a two-by-four that's one and one-half inches thick, you want a nail that's two and one-quarter to three inches long, or a 9 D to a 12 D nail. (There are no odd-numbered nails; the choice is between a 10 D and a 12 D. The 12 D does a better job.)

Plumbing Supplies

Plumbers either will try to talk you into using more expensive materials, thereby making a lot of money on them, or they will coax you into using cheaper materials, for which they can add on a hefty surcharge. Either way, they do well.

These days, it pays to make your drain lines out of *PVC* (polyvinylchloride) rather than copper. PVC works just as well and costs a fraction of the price of copper.

Obviously you should select your own

Sinks,

Tubs,

Toilets,

Faucets,

Bidets, and

Shower heads (although these can be changed easily and cheaply).

Let's talk a little about shower heads. Don't pick the water-saving ones (Oh, am I going to regret this remark later when my environmentalist friends read this paragraph). Many of the water-saving sprays have so little pressure, it's like showering in a mist. People often take longer showers with water-saving heads because it takes longer to get the shampoo out of their hair (and they end up using as much water as with a nonwater-saving shower head.)

Electrical Supplies

This is one category with which you don't have to spend much time fretting over materials. If you are using a basic system, most of the materials are standardized by the building code.

As long as your local inspector is going to examine the electrical work, you shouldn't have any problems with materials. Just make sure that the city inspector comes through for a rough-in and final inspection.

If you supply your own parts such as switches or smart-house equipment, make sure the electrician understands how those parts work. There are a lot of fun energy-saving and security devices available, but the electrician doesn't automatically know how these newfangled gadgets work. It's hard for the electrician to test an unfamiliar item. Keep the original instructions on hand, as well as the manufacturer's telephone number.

In most renovations, tradesmen prefer to tap power from existing circuits rather than to create new circuits. This is okay as long as the load is not too great. If you're adding a room to an existing circuit, tell the electrician what the room is going to be used for so that he can determine if the load will be sufficient. The preferred scenario is generally to run new circuits for all new areas. That way you have total control of the system you install. It's easier to troubleshoot for mistakes and problems later if an area is on a separate circuit. Try to coax the electrician into doing this.

You may need a *heavy-up* (a rewiring of your electrical system to support increased demand for energy) in order to make some of your new appliances work. If in doubt about whether your remodeled house is going to need more juice, lean toward adding more circuits. These days typical all-electric houses need between 150 and 200 amperes—more if you have electric heat. Call your local electric company for a (free!) calculation of loads. It's worth doing.

As far as fixtures go, electricians are generally happy to supply your fixtures at retail prices, even though they buy them whole-sale. They'll show you catalogs and price lists but not what they paid. If you are reimbursing the electrician for costs, make sure he shows you the paid invoices for fixtures, not the list prices—usually a 55 percent difference. This rule applies to almost all mechanical contractors.

Paint

The biggest decision regarding paint, after color, of course, is the choice between latex and oil paints. For lower-cost materials and ease of cleanup, latex is preferred. For long wear and good cover-age, oil is preferred. Oil also looks a little snazzier. The basic rule is that you can put latex over oil, but not oil over latex on reap-plications. Follow me so far? There are primers that allow oil over latex, however; but that's an additional expense.

As a homeowner, your choice of latex versus oil depends on both price and aesthetics. A good oil-paint job will cost 30 to 50 percent more than a latex paint. For most locations, latex is sufficient. If your house is already painted mostly with oil paint, then you might want to maintain oil throughout. A good compro-mise is to do walls and ceilings with latex but use oil on wood trims.

If you have plaster walls, make sure the painters use plaster on repairs, rather than *joint compound* (drywall mud).

Be sure to specify exactly which color you want where.

Flooring Materials and Finishes

If you are installing wood floors, you have to specify the grade of wood in conjunction with the type of wood, such as oak, stripped flooring, tongue-and-groove, two and one quarter-inch strip, random length, No. 2 common. Why? If you just say "oak" or "pine," you could get anything. You could get planks. You could get boards of short length that look like someone took apart a parquet floor. The parameters you must specify are

Type of wood,

How many knots are allowed in the wood (if any),

How long the pieces are,

How wide the pieces are, and

How the pieces fit together.

If you are going to create a design in wood, you had better specify exactly what you want and where you want it because the floor installer is never going to see the design in his mind the way that you do.

By the way, I don't recommend wood floors in kitchens or bathrooms. They look great—for a couple of weeks. This is really a bias of mine, but I think that wood floors in wet areas are too difficult to take care of and are to prone to discoloration, damage, and, ultimately, rotting.

Carpets

When you order carpet, you have to make sure not only that you get the right color but that you specify the texture and dimension of the padding underneath, the pile thickness, and the density of the actual carpet. These should be checked together at the show-room. A thick carpet does not need as much padding as a thin carpet to feel soft underfoot. When the carpet is delivered and installed, as with all other materials you must check that what you got is what you ordered. You have to make sure that the pile all lies in the same direction in your house so that when the installers unroll the carpet and cut for the rooms, they have preplanned that all the rooms will work together lying the same way. If you don't do this, the seams will show more so than if the carpet had been laid properly. Wall-to-wall carpets must be stretched, placed, and then cut at the walls so that the carpet doesn't stretch from usage, causing bubbles and waves (and precipitating a call to the carpet shop to reinstall the carpet).

Hire the Right Person for the Job Before Work Commences

One-stop shopping is convenient, less time-consuming than running around to different stores, and often cheaper. One contractor can sometimes be swifter, cheaper, and easier for you to command than five or six different crews performing specialized tasks.

But there's a downside to this philosophy. A contractor who specializes in one area—say, plumbing repair—may not be so hot in a closely related area such as bathroom remodeling. It may be hard to know this without the aid of hindsight because, although the tradesman's former clients may give him glowing marks in his particular craft—say, carpentry—they may have had no experience using that contractor as a plumber, painter, or draftsman. Lucky you: You can give that contractor his first opportunity to make big mistakes.

Alice Powers of Washington, D.C., experienced the menace of a specialist who branched into construction areas he shouldn't have. As Alice describes it, her family lived through the "bathroom from hell."

It started innocently enough when Alice's husband came to her one Sunday in 1988 and said, "I've found the house we're going to live in, and the best thing is that we don't have to do a thing to it." (This is the same refrain my wife and I sang when we found our 1905 house.)

This thought is typical of people who think they're never going to have to renovate their old house (the Powerses' house was built in 1923). The Powerses have now redone everything from light fixtures to the kitchen, one bathroom, and a deck. They had planned to renovate two bathrooms, but "the first was such a disaster."

The cause of the bathroom renovation was the first owner, who did nothing to the house for thirty-seven years, keeping it a "virgin house." Old houses like renovation now and then, especially the parts that get heavy use such as bathrooms and kitchens. Then came the person—a jack-of-all-trades kind of guy—who bought the house in 1960 (from whom the Powerses purchased it), who did his own renovations. While old houses like renovations, they don't like any old renovations: They need professionals to work on them. This former owner, according to Alice Powers, was handy but his craftsmanship was not according to code. She highlights this fact by adding, "This guy moved to Alaska, built his own house and two airplanes, both of which crashed." And so part of the Powerses' house crashed too: A piece of pipe from the third-floor bathroom broke and lodged between the second- and third-floor bathrooms. This is what inspired Alice Powers and family to renovate the bathroom.

They went about the project earnestly, first getting estimates that ranged from $6,000 to $30,000. They picked a bid in the middle, Aegean Plumbers at $13,000. After checking the Better Business Bureau, references, and a local consumer magazine—which all said that Aegean was a fine plumbing company—the Powerses moved on to the next step.

It started with the systematic destruction of their house.

Aegean said that it would take three weeks to renovate both bathrooms, and they'd be able to work in such a fashion as to ensure that one bathroom was always functioning.

Three months later, the first bathroom wasn't completed. But I'm getting ahead of the story.

"In one day, they ripped out the second-floor bathroom," Alice says, "and blocked the plumbing in the third floor," instantly putting both bathrooms out of commission for the duration. "They made the most colossal mess I've ever seen," she adds. "The five of us had no shower and no toilet. It was August."

Then things started to get bad. "The plumbers managed to create holes in the walls between the bathroom and the bedroom and zapped all the clothes in my closet and my husband's closet. Everything was coated with dust," Alice recalls. It seems that as the plumbers chiseled the tiles, they chiseled too energetically. "The dust was so terrible that we all had to move from the second floor to the third floor."

But these plumbers didn't restrict their damage to the immediate vicinity of the bathroom. They walked their pipes up the stairs without realizing that they would have to turn a corner. Of course, they walked the pipes directly into the wall," Alice says.

And how was the work Aegean Plumbers did? Was it worth waiting for?

No. "It took them four tries to get the tiled floor right," Alice laments. "I wanted white tiles; they put in beige. I pointed to the tiles and said, 'That's not white but beige.' They said, 'Yes, it's white.' I said, 'No.' They said, pointing to the tiles on the floor, 'This is white and the tiles on the wall are white-white. That's why the tiles on the floor look beige.' Then they put new tile on top of old tile. I told them to rip it up. The third time they tried to put the floor down, the subfloor was put in incorrectly. The fourth time they got it right but didn't have enough tiles. By that time, there weren't any more tiles in the store, so we had to wait another five weeks."

Imagine. "By this time," Alice says, "my kids hadn't had a bath in their bathtub for two and one-half months. My husband was showering at work." Alice had had enough. "I said 'Find these tiles and airfreight them!' " They did.

"The tiles arrived, but they put them in badly," she continues. "But I didn't care. Then the living room ceiling started to get wet, and it fell down. When I called Aegean and told them, they said, 'Are you sure it's not your imagination?' We had to put plastic down to protect the living room. Finally they came and fixed the overflow valve. But now there were cracks in the tile where the water seeped in. By this time, I had paid a $4,500 deposit."

Ready for this ladies and gentlemen of the reading audience? Alice says, "The plumber said he would leave it up to me how much I should pay. I was ready to pay half as the last of the work, putting in a radiator, was being done. Then the radiator exploded and covered the new tile with one inch of hot water. Meanwhile, no one was supposed to walk on the new tile for twenty-four hours!

"I decided not to pay any more. I figured that there would be repairs to do. And there were. I had $1,500 worth of plastering and painting. I had to hire other plumbers to come in to reset the sink. In August I noticed a crack in the tub. The tile around the radiator has cracked in a ray-formation.

"Aegean Plumbers has been back to do grouting, caulking, fixing the living room ceiling, and to look at the crack in the tub, which they say is the fault of the tub's manufacturer."

There's more, according to Alice: "They had never cut in the base tiles. They had simply glued the tiles onto the existing wall tiles. There were little things too. We had had an outlet near the mirror for the electric razor; the plumbers moved it to the center of the wall.

"The frame of the window hangs out wrong—it's recessed at the bottom and extends outward at the top. The sinks were installed ugly. They used plastic piping and made no attempt to make the pipes blend in with the tiles."

With a sigh, Alice sums it up, "These guys weren't heavy into aesthetics."

The plumbers, who had arrived August 2, announced on November 15 that they were ready to start on the second bathroom. Alice says, "My kids begged me not to let them."

There's practically no way to predict such disasters in advance. Alice Powers did everything correctly—checked references, interviewed the head of the company—everything. When you're doing a bathroom, she notes, "you're held hostage. I put a big deposit down. They kept reassuring me that things would get better." Even if she had wanted to switch contractors in the middle, most contractors won't pick up somebody else's work. If something goes wrong later, whose warranty covers the work? The first contractor's or the second's? Besides, the inspectors will only approve the original plumber's work.

Perhaps the only two changes she could have made were to insure that the plumber subcontracted to specialists for particular tasks such as tiling and window installation and that there was a general contractor overseeing the project and directing the workers. Alternatively, she could have been her own general contractor (as she was for her kitchen, which went well), but the decision to be your own full-time general contractor must be made at the onset of the work, not as things go wrong. Workers listen to a general contractor who is with them from the beginning, even if it's a homeowner. Later on you won't seem like a general contractor; rather, you'll sound like a complainer. Second, as Alice herself notes, she could have "put an end date in the contract and made being paid contingent upon meeting that date."

Hindsight is one of those special powers that you acquire during home renovation.

The aphorism, "Hire the right person for the job before work commences," isn't universally true. This rule works best for smaller jobs. For bigger ones you may have to hold off until the start of a particular component of the renovation before hiring the required individual. For a five-month project beginning June 1, you simply can't say to the plumber, "Okay, we'll be installing the second-floor bathroom fixtures on September 22." Even Jeanne Dixon won't have that information from the outset.

Repair Renovation

A large number of renovations that take place each year are renovations that homeowners must do because something has broken: A bathroom leaks into the kitchen below; suddenly, one spring, termites invade the house; the electrical wiring needs to be replaced; a pregnancy necessitates removing lead pipes; the floors are sagging a bit too precariously; or water drips in through the roof. A house is an organism; parts wear out and need to be repaired or replaced over the years. No house is immune.

Repair renovations don't allow much time to plan. The first step you should take is to contain the problem. If a bathroom is leaking, plug the leak or don't use the bathroom. If a beam is sagging, remove any weight from above the beam and support it from below. If you can, put some time between the onset of the problem and beginning to fix it. Even emergency renovations should be thought out so that, if you want to, you can do other remodeling at the same time. Turn a problem into an opportunity.

Who should you select to do a repair-renovation? If you have a serious emergency—a burst pipe, leaky faucet, failing heating system in February—you must tend to that problem immediately. But do not feel obligated to use the same person who handled the emergency work for additional remodeling, especially if you found the emergency tradesman in the Yellow Pages. While you would encourage treatment by a paramedic on your way to the hospital after a heart attack, you would prefer that a surgeon perform the coronary bypass. In the same way, halt the emergency, then select your contractor with the same intelligence you would use if you were starting a nonrepair renovation.

In many cases, for structural work the best person to start with is a structural engineer or builder. Alternatively, begin with someone who has a bright flashlight, useful tools, and excellent eyesight who has also done this kind of work before.

There are three parts to a repair-renovation: first, the emergency repair; second, investigating whether there is any hidden, related damage; and third, remodeling.

If you are undertaking a renovation to repair damaged parts of your house, it's essential to make a close, invasive inspection of those areas of the house. Even professional inspectors can overlook termite, water, and other types of damage. You may never know how extensive your repairs need to be until an exposed inspection is undertaken. Don't just repair visible damage. Don't self-diagnose the problem. Things must be taken apart: Pull up floorboards, remove pieces of plaster, squeeze into crawl spaces, get on top of roofs, slither under porches, and move laundry machines and other appliances. This is the kind of inspection that will tell you how extensive a problem is.

Again, the person who performs the emergency repair shouldn't necessarily be the one who investigates for further damage.

Don't only have a plumber look for problems surrounding a bathroom that is falling into the room below. The plumber isn't trained to find dry rot in the joists beneath the bathroom and he isn't going to be paid to fix it; in fact, spotting dry rot may delay and complicate his work. When in doubt, hire a structural engineer, a home inspector, or a builder whom you trust to thoroughly diagnose your house.

Visual inspections can reveal hints that something is wrong, but as a rule of thumb, only exposed inspections are diagnostic. This is especially true with water-sodden areas: They must be deeply exposed.

The amount of money you spend on an exposed physical inspection and repairs (and putting things back together again) will save hundreds or thousands of dollars later, money that you might have to spend ripping out all this new construction to get at a previously undetected underlying problem.

In fact, if you believe in rationalization, here's a good one: Remodeling after an emergency repair often necessitates that you tear things apart. If you're going to wreck a part of your house, it might as well be for a worthwhile cause.

A homeowner in Washington, D.C., bought a Victorian house and had the usual physical and termite inspections. The inspector looked closely but exposed nothing, which is typical for home inspections. When Paul Locher was performing a structural inspection for a renovation to convert a bathroom into a breakfast nook that the homeowner was planning, Paul noticed that the floorboards in the bathroom had begun to rot. Paul suggested tearing up the floor to see exactly how rotten they were, since they were over a hard-to-get-to crawl space. What Paul discovered was about 300 square feet of termite-damaged joists in a three-story load-bearing wall. The termite inspector would have found an indication of this, had he ventured into the crawl space beneath the bathroom, but he didn't. Crawl spaces are yucky. The termite damage cost about $3,000 to repair. If that sounds like a lot, consider that if the problem had been detected after the breakfast nook was finished, the repair would have cost nearly $10,000.

Take the case of the McLean, Virginia, home shopper. When the potential buyer's builder (not the inspector) went to scope out the house for renovation, he found signs of water leakage in the attic. Only a little bit, but enough to raise eyebrows. There were other caution signs: cobwebs with dead insects and mouse traps. A virtual Twilight Zone animal kingdom, the builder said.

The builder wanted to do an exposed inspection of the attic, but the seller said, "No, it's nothing a little caulk won't cure." The seller wasn't going to allow the attic to be ripped up before the sale. Finally, the seller relented and allowed some small areas to be disassembled. The builder found termites and carpenter ants that had damaged sill plates and roof rafters. Obviously, there was more to fix in this attic than merely dampness to caulk.

An exposed inspection is necessary to reveal the extent of damage, but the inspection doesn't necessarily tell you the cause. Indeed, it can be expensive to leap to conclusions about causes. The quick assumption in the McLean, Virginia, house would be that the roof leaked, but that would have been a costly assumption, because water had actually entered the attic as a result of clogged gutters.

If you see a small problem, especially in a structural area, and especially if it involves water, assume that there are serious problems lurking.

Once the Work Starts, Don't Be Too Patient

"We were much too patient. We like to camp and we don't mind budget accommodations. We put up with things longer than we should have. We should really have put our foot down," Michele Sands says. Don't be a therapist to your contractors. Her husband adds, "We both tend to overcompensate psychologically. If someone is having problems or making mistakes, we tend to jump in and say, 'How can I help? What can I do?' And they will take total advantage of that."

Allow the work to proceed according to the plan and allow for unexpected changes in the plan, but don't let the plan continue through eternity. If you don't keep reminding workers that the plan calls for a deadline, they will never finish. Why should they? Every week they show up, someone gives them a check, so their main incentive is to continue showing up.

Inspect All New Fixtures and Appliances

Inspect the appliances you are going to install, especially any parts that will become inaccessible when installed.

When our Jacuzzi arrived it looked like it was in order, but our builder, Paul Locher, decided not to take any chances. After all, the boat-size tub was going to be placed against the outside bathroom walls (not an easy feat), and once there, the backside would be inaccessible. The side of the Jacuzzi that was going to go against the wall held the vacuum lines for the control switches—the switches that let you turn the jets on and off. As it turned out, the switches were not installed correctly at the factory, and Paul had to reattach them and make sure they had a good seal. Later, the only way to get to the switches would have been to cut through the exterior wall on the second floor of the house.

When you buy a new car, it's expected that there are going to be bugs, and you will have to bring it in for a checkup in a few thousand miles. Unfortunately, home appliances and fixtures can't readily be disassembled and shipped back to the manufacturer.

In the case of our Jacuzzi, a one-minute check-over and five minutes of work kept the Jacuzzi functioning. I would have hated to get someone to drill through the wall or, worse, try and get the Jacuzzi back out through the windows.

Whether you are your own general contractor or whether you hire a G.C., learn to notice. Examine everything that comes into your house. If a material or appliance doesn't look right, it probably isn't. Turn knobs, open doors, flick switches, inspect seals and connections, gently tug on wires, examine screws. A door that's not perfectly even should be planed before you install it, not afterward.

Do you have a carpenter's level handy? Use your level to check all new doorframes, floors, stairs, shelves, windows, railings, and other surfaces to see that they are straight and level. A level is the most useful tool you have to check that work is being done correctly. Often shoddy work first manifests itself by not being level. Insist that anything crooked be rehung immediately; otherwise, be prepared to put your furniture on shims or have a slight seasick feeling when you walk through your house.

A tape measure is your second line of defense. Verify that rooms, alcoves, spaces for cabinets, showers, sinks, and other areas are being made according to the dimensions specified in the plan. A few inches here, a foot there won't bother the builder, but it will make a difference to you when you move in that baby grand piano.

Working in the Winter

Winter construction takes special planning. Cold-weather work often costs more than summer renovation, but it doesn't have to be too expensive. Workers want to be kept warm and the energy to do that costs money. A plastic envelope should be placed over the work area if it's exposed to the elements and propane or electric heaters installed. This is sufficient to enable any kind of work to be performed. The upside to working in the winter, however, is that the colder it is, the faster materials can be obtained, since there's little competition.

Don't worry, ma'am. It won't make much dust.

Plaque given to Penny and Don Moser after their renovation

8 | Preventing the Destruction of Your Floors, Antiques, Appliances, Lawns, and Other Valued Possessions

WE HAD TO PROTECT OUR FLOORS ourselves with rubber mats, even though the company was supposed to do it," says Michele Sands. And that's the way it is. Contractors have a Genghis Khan philosophy toward your personal possessions. They treat every item in your house with the attitude that most of your things belong to them or in the garbage.

And they believe, truly, that the more dust they create, the more it looks like they're working hard.

The Daily Inspection

"Unless you are very lucky you are going to have to inspect each detail," Harry Hobkin discovered. "You are going to have to understand exactly how to read the blueprints."

In detail and frequently. You should inspect the job every day. No, not should. Must. Absolutely *at least* once a day. Inspecting the job every hour is even better, but of course that tends to produce anxiety and a lot of lost job time. Still, the more often you inspect the wonderful and terrible things that are happening to your house, the better the final result will be. It's not unusual to spot a half dozen things gone wrong during each inspection. Catching mistakes before they become permanent will save you from eternal grief. It takes only a few minutes to make a mistake

145

that is not necessarily irrevocable but a real challenge and chore to change nevertheless. For example, the following tasks take less than fifteen minutes to do, and will take hours to repair:

Windows installed in the wrong place,

Electrical outlets cut in the wrong place,

Using lead solder on pipes instead of silver solder,

Gluing the wrong thickness of subfloor, and

Making a room one foot too short.

Contractors have a different point of view regarding your inspections. Builders want their work to look permanent, final, irrevocable. Contractors count on the fact that when something is done, even if it's not the way you want it, you won't insist on the work being redone. They look at your inspections as tours of approval.

Assuming that you can't be on the job every hour, the best time to inspect is first thing every morning when the tradesmen start work, so that if you spot something wrong, the workers have all day to correct it. It doesn't hurt to ask the workers—politely and intelligently—what they'll be doing that day, what parts they're waiting for, have there been any surprises, do they need anything to make their jobs easier. As you ask questions, you don't want to sound too ignorant about what's going on. If you appear . . . well, stupid, the contractors are going to take advantage of your ignorance. For example, if your question indicates that you have no idea that the joists are about to be placed on top of the foundation wall, the carpenter knows you don't know the first thing about construction framing. Everybody knows that the joists are the next thing to go on top of the foundation wall! (Now everyone does.) If you can't ask an informed specific question, ask a more general one.

Talk Is Cheap

Always walk a new contractor through his whole job prior to the start of any work. Make sure you and he understand the scope of the work in the same detail. That way you get an education from him without giving away your ignorance, and he gets to listen to your concerns, whether he wants to or not. You might also pose "what if" questions to the contractor about particular problems they might encounter while working. This is the time to turn on your fiddlesticks detector. If the answers you get during this prework walk are tentative—"Gee, we won't know till we tear the wall up"—ask the tradesman to call his boss, or you call his boss before the work starts. You want specific answers. Have the workers show you exactly what they're talking about. For example, if the HVAC guy says, "You can run a duct through this wall and along the floor to this endpoint," have him mark exactly where he's coming through the wall and which floor joist he's going to be meeting. Go to the other side of the wall and see what he's about to take out—electrical lines, a pipe, perhaps the space where your bed goes. Undoubtedly there will be alternative approaches. It's far, far better for you to pick from these alternatives than for the HVAC guy to decide. Believe me, it's much better. The HVAC workers don't know what your living style is. "They won't see it from *their* living room once the job is done," says Paul Locher. And, as long as the equipment works, they won't care about the fine details either. This goes for all work. The theory, according to contractors, is that if it works, you're satisfied and so are they.

It's important that the contractors know you are going to be there every day. They need to know that you are going to make them explain what they are going to do, and that you will make decisions.

Every contractor loves beating up on other contractors. They love blaming the people who went before them. You may hear the carpenter cry, "Damn mason gives me a foundation out of square, and six inches longer than the lumber I wanted to use." Or you may hear the plumber say, "Damn carpenter framed this, and I can't put my vent stack right behind the toilet." Or, the electrician may say, "Damn plumber. I can't put my switch box on this wall because the plumbing lines come through here." Or the drywaller will say, "Damn electrician, didn't drop his boxes out of the studs a half inch for me to drywall to."

The translation is, "Don't blame me if you don't exactly like the way the work comes out aesthetically." The interpretation of this translation is, "I don't really care what things look like, anyway."

What you have to insist on is that the work come out exactly the way you want it, or at least only microscopically different from your original expectations. For example, let's say the electrician is having difficulty putting in an outlet because a preexisting structure, in this case a pipe, is in the way. He wants the plumber to move the pipe. Instead, tell the electrician to move the electrical outlet to the next stud, fourteen inches away, instead of moving the plumbing lines or not putting in an outlet at all.

Before Everyone Goes Home

When workers leave your house, that's what they do: leave. Workers don't have that sense of dread that overcomes homeowners as they're driving down the interstate and think, "Did I lock the doors?"

If you're lucky, 75 percent of the time the workers will remember to lock your doors if they leave before you get home. If you're lucky.

But not locking a front door is a very, very minor problem compared to the other things they can forget to do. A couple was building a new basement under a brick row house in New Orleans. Each day, the contractor hired day laborers at the downtown employment stations (places where people looking for work hang out). The quality of these workers was on par with the caliber of most state legislators—you wouldn't want to trust them with anything too scientific. Their task was to dig out an entirely new basement under the row house where no basement had been before. Digging is tough work and there's a special way you're supposed to do this so that an interval of support remains as you're creating the space. One Friday these guys had completely dug out one corner of the house and for the weekend. The owners didn't notice, and probably didn't even know what to look for. To secure the floor above, the workers placed a six-by-six piece of wood under the house. The top of the post was under the corner of the house. The basement beneath was just raw earth. Then it started raining. That was Thanksgiving evening, and the owners' entire family had gathered in the house over that part of the basement. Rain got in the basement. Suddenly, the dining room began a slow descent toward the basement—it felt like an earthquake. The entire room started to shake and slip into the pit beneath. Fortunately, the Thanksgiving party managed to get the wall propped up. But by then, the builder-excavator was long gone for the holiday. As with many of the tales of tragedy in this book, there are a number of morals to this story, but one is paramount: If crucial work is being done, don't let the workers leave until potential problems have been eliminated.

Oh, what the heck. Here's another moral: Be especially suspicious of work that's just completed before a holiday or long weekend.

Here's a list of things you should check before the workers go home for the evening, and especially before they go home for the weekend or on vacation:

If your roof is being repaired, is there a secured tarp on top? (Do you have buckets and trash cans around, just in case?)

If your gutters are off the roof and there is the slightest chance of rain, get those gutters back on or everything in your basement may be destroyed. As you probably know, removing gutters can actually cause rain.

If plumbers have been around, make sure they've left you with a working toilet for over the weekend.

Check your grading.

Check that the house is secure (a locked door is no good if there's a whole wall missing).

Do you have heat?

Is the electricity working?

Are the toilets and showers functioning?

Roofs and Rain

Intellectually, I've always wondered, How do they keep the rain out of a house when they take the roof off to repair it or to lower in a grand piano. If you've ever looked at a house where this is happening, you'll get the answer: They put a giant baggie on top. A clever idea, but unfortunately, plastic is not as sturdy as shingles, and you're guaranteed a leak. Or twenty, as happened to one Virginia homeowner. She eventually had to place a dozen or so glasses, cans, pots, and other concave objects in various random positions under her house's baggie.

A giant baggie presents other problems as well. Many houses, you'll notice, have angled roofs. This tells the rain in which direction it should fall. A baggie sends no such signal to the rain, especially since the position of these tarps is adjusted by the weather. Consequently, the water's weight tries to pull the entire thing down into the house. Mark Richardson, a builder, says, "Once there was so much water in the tarp we had to pump it out."

Three quick lessons here: First, Have myriad cans around for when it rains; Second, Monitor the integrity of the tarp daily; and Third, Be prepared to order your contractor to find a water pump. On this score, don't accept a scratch on the head that says, "I've never heard of that before." Insist.

What If You Are a Perfectionist?

Which aspects of remodeling can be made perfect and which can't. Some elements of renovation are more prone to error than others. Your limitations in striving for perfection are materials and money. Surprisingly, no amount of money will ensure perfect work. The finer the materials, the more you are going to notice the most minute scratch. If you have a pine cabinet with a scratch, you are less inclined to be upset than if the cabinet is made of mahogany.

However, you have a right to better workers when you build with finer materials. If you've noticed that the workers' skills are a par or two below average, don't let them handle expensive materials and fixtures. Many don't "understand" quality. So-so plumbers installing a $500 sink are far more likely to scratch it, or to install it in such a way that oodles of caulk are needed between the sink and the wall, than are better-trained plumbers. A worker who's used to working with ordinary materials can probably do no better than ordinary work. Let me put this another way: The quality of what you buy is only as good as the person installing or constructing it.

Your Daily Energy Bill

Home renovation is expensive. Just about the only thing in the process that doesn't cost much money is this book. Everything else—labor, materials, architectural drawings, hauling—is fantastically expensive. There's also another cost that can excruciatingly—and surprisingly—add to the cost of renovation: energy. Specifically, gas and electricity. Someone's got to pay for all that juice, and it's undoubtedly going to be you. There's really not much you can do about the amount of energy contractors use while on the job, but there are some steps you should take to conserve energy that is not specifically being used for work. These include not running your air conditioner during renovation. If you feel guilty about the workers sweating in ninety-degree weather, keep in mind that they are used to it. They expect to work in hot conditions.

There are other good reasons not to run your air conditioner during renovation. With so many windows and doors open, you'll simply be cooling the outdoors. And it is important that windows and doors be open during remodeling because it's easier for dust to exit when they're open. This brings me to another consideration: The dust that's kicked up will get into your air conditioner and clog the filter in a couple of days or, worse, damage the system. This is especially true for central air-conditioning systems, which tend to be very expensive. Finally, air conditioners make noise, and the extra noise makes it hard for workers to communicate among themselves.

> **Insist that all lights be turned off at the end of the day (if no one is living in the house), or that the lights be turned off in the part of the house that's being worked on.**

This simple rule is among the most widely ignored. Workers couldn't care less about the cost of electricity to you and aren't about to climb over debris to get to the light switch . . . unless you ask politely. The best way to do this is to leave a note.*

*I know, you're thinking, "But Adler said many workers can't read or can't read English, so what's the point of leaving a note?" The note will be read by those workers who can read, and as for the others—well, lights left on is not a really big problem. You do what you can.

Leaving lights on is not the most major of issues you're going to face, so you shouldn't make a big stink, but reminding the contractor lets him know that you are cost conscious. As for worrying about vandals thinking no one is home if all the lights are off, between the dumpster and all the people coming in and out of your house, leaving a light on is not going to convince any would-be burglar that your house is occupied.

Keep the heat down in the winter. Not too cold, because concrete, paint, plaster, polyurethane, and other materials need minimum temperatures to cure. And you don't want to take the chance of your pipes freezing. But there's no reason either to keep your new hardwood floors comfortable all by themselves.

Plan your construction so that you lose the minimum amount of heat. Take the case of Michael Weiss and Phyllis Stanger. Their new addition was coming along perfectly. The walls were built on time, the skylights were placed properly, and the floor ducts were installed to perfection. But the ductwork to their addition was connected to the rest of the house so that it could be controlled by the central heating system, and this created their problem. Because the new room's heating system was interlocked with the rest of the house, during the frigid month of December, as Michael and Phyllis heated their house, they also pumped heat into the unfinished addition. They were essentially heating the yard. Try to schedule the work so that the insulation comes into the new room around the time the ducts are connected with the rest of the house. The problem in Michael and Phyllis's case was that the windows couldn't be put in because the window manufacturer shut down for a two-week vacation in December. The lesson of this particular story is: Plan to have all your materials ahead of time. If you have to leave a wall or window unfinished, seal the ducts in the addition by covering them with plywood or sheet metal, or wait as long as possible before making the final connections between the addition and the main house's HVAC system.

Close the windows at the end of every day. This keeps heat in and rain out. This simple formula comes highly recommended not only by energy-efficiency experts, but by everyone who's had rain come in an open window and ruin a newly laid wood floor. If there are no windows yet, place tightly sealed plastic over the window openings. Every night. "Tropical storm Cliff came in with six inches of rain," Penny Moser says of her open addition that was under construction. "We would be out there at night with sheets of very heavy plastic trying to keep water out." Keeping windows closed also helps keep moths and other egg-laying insects from taking up tenancy in your house. A New Jersey homeowner had birds fly in through open windows.

When the general contractor isn't on the job, don't count on subcontractors to cover your windows, roofs, and doors. It's not their department to protect your house.

The Case of the Missing Light Bulbs

It's not actually a mystery, though it may seem that way at first. Where have all the light bulbs gone? You may notice that bulbs are disappearing from fixtures at an alarming rate, with no apparent explanation. The bulbs aren't dying, they're disappearing. Where to?

Well, if there are painters around, you can reasonably suspect that the painters are using them in their *droplights* (cheap spotlights). Now, it's a known fact* that painters drop their droplights frequently, something you might expect because they have to prop the light at complex angles to get a good view of what they're doing. When the droplight drops and the bulb breaks, the painter simply gets another bulb from wherever is handy. You may not like this process, but there's nothing you can do that will change it. I suggest buying a case of the cheapest seventy-five-watt bulbs you can find. At least now you know that some elf isn't stealing the bulbs.

Protecting Your Loved Ones (TVs, Couches, Paintings, Etc.)

The most effective way to protect your personal belongings is to move them out. Put them in storage. Loan them

---

Known facts are those facts which are true regardless of anyone's denials. For example, when the floor installer swears that he locked the front door, which was unlocked throughout the night, but he was the last person to leave the house, it is a known fact that he did not lock it.

to a museum, your sister, your neighbor. Get them stolen; they'll certainly be safer than in your house. No one can inflict more damage on your possessions than subcontractors. When reincarnated subcontractors come back, they return as airline baggage handlers.

Hire a moving company to shift your things into storage, and you won't have to worry for a second or pop a single Mylanta over the possible damage that could happen. (Be sure to take out insurance for the moving and storage too.) This should be one of the first steps you take in the planning process.

ITEMS TO GET OUT OF THE HOUSE OR INTO A PROTECTED AREA

Jewelry

Crystal

Silver

Electronics

Tools

Antique anything

Art

Chandeliers

Personal items such as photo albums and keepsakes

Rugs

Furniture you don't want marred

Clocks

Computer disks (not to mention the computer)

Your renovation contract and plans

Once all your precious (precious being a subjective term) belongings are out of the house, mark a path where you want workers to walk. Use a plastic walkway, drywall, paper, or canvas to identify the path. If you have sawhorses and "Police Line—Do Not Cross" tape, use that. There's no absolute guarantee that workers will keep to the path, but I can assure you that they'll walk wherever they want if you don't put a path down. Make the path look like a path, however. Once upon a time I owned a white rug. Because I didn't want workers walking over the rug with their muddy boots, I placed an old blanket over the rug to form a path. In the most conscientious fashion, two subcontractors walked single file around the blanket and on the rug.

Another more radical step is to construct a partition to keep workers out of certain areas. Drywall partitions are not expensive to build and are worth every nickel.

Protecting Floors

Put paper, plastic, or (if you're especially protective) drywall on any floor you want to protect. If you use paper, two layers is a minimum. A special type of paper called rosin paper that comes in giant rolls is available at many hardware stores. Don't use ordinary paper because it won't last. If you use plastic, put paper on top because plastic is too slippery a substance to work on. Just because you've put down a protective layer, don't assume that your floors are safe. These protective materials have a way of vanishing, by magic, by overuse, and by displacement when they get in the way of workers. Workers don't like floor protectors. Keep an extra roll of rosin paper on hand.

Remember the On Switch

Not everything that goes wrong is the contractor's fault. One homeowner in Washington, D.C., had a two-zone HVAC system; a furnace and air conditioner in the attic and another in the basement. Throughout the winter, she heated her house and complained to the contractor that the upstairs wasn't warm enough. She wanted him to come out and balance the two systems. The contractor decided to come out while everything was still under warranty but couldn't seem to understand the problem because everything was working fine when he left. He came into the house, turned on the downstairs system, went upstairs, and found that the upstairs furnace had never been turned on. At least she saved on her heating bill.

Ladders

Ladders are dangerous, not to the person working on them, necessarily, but to your home. Ladders are unwieldy to carry, and workers think nothing of banging them into your chandeliers (which should have been removed before the work started), staircase railings, doorframes, paintings, refrigerator doors, or piano. A worker will not think twice about putting a ladder that's just been outside and is caked with mud on top of an expensive rug.

Always demand that the subcontractors place cloth pads on the tops of ladders because ladders also mar the surfaces they lean against. Same thing for the ladder's feet: Forbid any ladders that aren't protected on the bottom with rubber or cloth. Otherwise, at every spot where a ladder has been you will have two little marks on the floor and two little marks against the wall. At one Pennsylvania construction project the electrician had come in to install the exterior light fixtures in a house with stucco walls. Everywhere his ladder leaned against the wall were two indentations in the stucco. Wherever the ladder goes, expect marks.

One painting company I learned about had their painters use stilts instead of ladders. Some drywallers I met used these two-foot stilts, too. Not only did this prevent unsightly ladder marks, but the painters could get around swiftly without having to move their ladders every few feet.

Protecting Lawns

Actually, I should title this section "Minimizing Damage to Your Lawn." There's not a whole lot you can do to save a lawn that's a pathway for heavy machinery. Equipment such as backhoes that go across your lawn and appliances that are dragged across the lawn (the new refrigerator, still in its box) will kill your grass. Boots going across the lawn a dozen or so times a day will severely damage your grass. You could try to place protective material on the lawn, but inasmuch as grass needs air, sunshine, and water, this will kill your lawn too. Your best bet is to clearly mark a path for equipment, appliances, and workers to travel. (Workers, like birds, choose the shortest path to travel, which is not necessarily your front walkway.) Make the path crystal clear! Use police ribbon, plastic cones, or barbed wire. Better still, put a clause in your contract that says the subcontractors will obey the path or the cost of replacing the lawn will be deducted from the G.C.'s pay. Of the above-mentioned physical barriers, plastic cones seem to be highly effective. Something from our early school-crossing experience makes us all want to obey them.

Also make clear to your builder that he should not store trash or other materials willy-nilly on the lawn. Debris tossed from the second floor should go down a chute into a dumpster, not on top of your azalea bushes.

Good luck with all of this. Lawns don't really stand much of a chance against builders, but you may be lucky.

Outside

Every time I walk out our eighty-six-year-old Victorian house, the first sight I'm treated to is the tree our painters cleaned their brushes against. Gets the blood boiling. They did it in the backyard too, so there's no salvation there, either.

However, after I visited Katie Jacobs and Mark Halloway's house in Virginia, I felt much better about our idiotic painters. After all, at least Peggy and I didn't have our entire lawn destroyed.

"If you don't put up an absolute do-not-cross barrier, they'll store stuff there, they'll track mud there, they'll dig there," Mark says. Katie and Mark pointed to where a three-foot mound of dirt leaned against the front of their house, a remnant from an excavation in the back to build an addition. "I thought that night, 'Tell them in the morning they have to get that dirt off there,' but I was too busy with ten other things and I just didn't say it. We should have, right then, or put up a barrier beforehand. They were walking back and forth over the newly planted trees, through the bushes. They were walking on our neighbor's property."

The digging in the back of Katie and Mark's house required a small bulldozer and other digging tools. "The contractor told us that all they would need would be a path ten feet wide. The builder assured us, 'No machine will be bigger than that. You wouldn't be able to get in the back anyway.' "

It turned out that the contractor had to take down several trees to get the backhoe/bulldozer into the back of the Jacobs-Halloway house. " 'We'll be responsible for the trees,' the contractor said. I should have gotten that in writing," Mark reflects.

Back to the front of the house: Katie and Mark's beautiful expanse of lawn became so damaged, so dangerously damaged, in fact, that their postman refused to bring their mail. He brought mail to their neighbors, who had vicious dogs, but not to the Jacobs-Halloway house. Finally they created a separate path for the mailman along the side of the house.

The Jacobs-Halloway walkway was now composed of odd-shaped pieces of plywood. Unfortunately for Katie and Mark, their tale of woe didn't end there. Sometimes, as they discovered, you have to fight, fight, fight with the contractor to keep him from destroying your property. After excavating a sufficiently large hole plus a drainage area, the backhoe/bulldozer was left with no room in which to move forward and return to its place of origin, the street. Mark says, "One night we got a call from the sales rep. He said, 'We got the Bobcat [digging vehicle] out, but the backhoe is still in the backyard and we want to take it around the other side of the house.' They recommended cutting down the dogwood tree. He said, 'Don't worry, I can replace it for a thousand dollars.' We said, 'Absolutely not.' First of all, it would be across the neighbor's property line, and we know a few things about trees. Those other roots are vulnerable. We had said that that side of the house was not an access route. That was clearly established. But the backhoe was dug into a corner. They had finished the excavation with sharp walls exactly where we thought the foundation walls were going to go. There was no way to get that thing out of there. It was dug into a pit. So we argued and talked. They said, 'Can't we just put down some plywood?' We said, 'No, that's still a lot of weight over the tree roots.' I called the Forest Institute to ask them what could be done. The Institute said, 'You could put down bales of hay along the entire route with plywood over the top. But you'd probably lose the dogwood tree.' We thought, Why take the risk? They had screwed this up."

Katie adds, "What they wanted to do was what was most convenient for them."

Then the sales rep told Katie and Mark, "We can't go around that way, so we've decided we're going to rent a crane and lift the backhoe over the top of the house." Mark says, "I almost laughed. I was stunned with disbelief. I asked, 'What if the cable breaks?' He said, 'Then we've bought your house.'

"So we said no to that," Mark adds. "This was all making us very nervous. The next morning the foreman showed up early, and by himself drove the backhoe over a very narrow strip on the dirt-path side of the house. The backhoe could have toppled, and he would have been crushed to death. No one had really planned the excavation."

Excavation

Despite the brute force required, excavation takes a delicate touch. Doing an excavation for a foundation isn't just digging a hole in the ground. Thoughtful preparation is absolutely necessary. Here are some guidelines:

1. Carefully check for all utilities on the property. Every contractor hits utilities. One group of subcontractors was digging in Virginia and using the power company's maps to find where the utility lines were, but as soon as they started digging they cut the water and sewer lines. These lines were not properly located on the map.

2. The least-thought-out product of excavation is dirt and where to put it. Your garage may be coveted by the contractor, but that's not the place. It doesn't make sense to take the dirt off the site and then bring in more dirt to backfill later. But if there's no place to put the soil, you may not have a choice. Excavated soil should not cover tree roots inside their drip lines, should not block the concrete truck's access to the site, or prevent other materials from being delivered to the site. You might also want your mailman to be able to get to your front door.

3. Trees are tempting targets for earth-moving machines. Running over a tree's feeder root can quickly kill a tree.

4. If the concrete is going into footings (poured concrete foundation walls), plan ahead and have the contractor build ramps so the concrete trucks drive up and over the foundation wall and pour right into it directly from the truck. The cost of the concrete pouring will increase if the mixture is pumped to your house from the truck in the street via a pumper crane.

5. Don't let your backfill dirt pile mix with the trash pile. The two materials cannot be used interchangeably, despite the virtues of recycling, and some subcontractor will try to use trash as backfill. If you have to start moving the trash pile with the machine, you end up tripling the weight because dirt is moved into it. As trash disposal is based on weight, the cost of removing the garbage will increase. This is a very, very common problem. It's a result of laziness, sloppiness, and poor management.

What I like best is the bit where they cut off your water and electricity, and food and oxygen supply, and rip your house into tiny, Chiclet-sized pieces and announce, "We'll be back on Thursday," knowing that's the last time you'll ever see them.

Dave Barry on subcontractors, *Metropolitan Home,* October 1990

9 | The People Who Do the Work

What They're Supposed to Do, Why They Don't Want to Do It, and How to Make Them Do It

HERE IS A SHORT LIST of some of the people you'll meet during your renovation. Some will bring joy and happiness to your house (in the long run), others will seem to serve no purpose at all, and still others will seem to have the single-minded function of making your life miserable.

Demolition "experts"

Concrete pourer

Stonemason

Carpenter

Electrician

Plumber

HVAC installer

Insulator

Drywaller

Painter

Roofer

Excavator

Tiler

Carpet installer

There's some valuable information in this chapter, perhaps the most valuable information in the book. Because home renovation is a labor-intensive process, the people who work on your house determine how well it comes out. But it's more than just the pure technical skills of the people involved that make a difference in the quality of your renovation; how much energy, interest, and devotion they put into your renovation is crucial. This chapter tells you what each person—carpenter, concrete pourer, electrician, and so forth—should be doing, how to tell when they're not working up to the highest standard, and how to make them do their best.

> Remember, in many parts of the country there's a
> shortage of skilled labor. Any given person who works on
> your house may not be highly skilled but a trainee. This is
> especially true when there is a lot of construction or
> renovation going on in your area. It's also true when the
> construction business is in a slump: During those times,
> skilled workers are laid off and part-time workers are
> hired to do your job.

Even with renovation, it's buyer beware. "I went into our renovation thinking that the workers were going to be professional," says Julie Johnson. "You need to keep them responsible by keeping a close eye on them."

General Observations about the Tradespeople Who Work in Your House

If there's one single piece of information you ought to know about how tradespeople work, it is this: Each subcontractor cares only about his particular job. If his work is done (note that I did not say done correctly, merely done) and he's paid, then he's satisfied. For example, the painters who painted our house didn't use drop cloths. The floors, they were told, were going to be sanded and polished, so there was no need to protect them. What they didn't know was that not all the floors were going to be sanded, though all were to be polished (and the staircase banisters—now dotted with white specks—weren't going to be touched at all). So now we have some wonderfully shiny pine floors with little white flecks on them. The same painters (oh, how I love them) did an excellent job of painting the kitchen. However, while painting the kitchen, one of them balanced himself on the ladder by holding the doorframe in the dining room, anointing the dining room doorframe with hand prints. But as long as they didn't get fingerprints on their handiwork, they were satisfied.

Welcome to the Land of 1,000 Excuses

When Penny and Don Moser created a deck on top of their two-story addition, they wanted a door between their second-story bedroom and the deck because their bathtub was there. The builder took measurements to turn a window into French doors but didn't get the measurements exactly right. When the doors were installed, there was a four-inch gap between the bottom of the door and the doorframe. Penny pointed out the gap and asked the builder what he was going to do, to which the builder responded that he was going to put in an extra step. Penny said that she didn't want a step, that it would be easier to get out to the deck without one. Ah, the builder said, you'll want the door raised because of the snow. The snow? That's right, he said, when the snow comes you'll want the step there to get you over the snow. Sounded reasonable until Penny realized that Washington, D.C., is in the South, and the number of significant snowfalls is small.

Welcome to the land of 1,000 excuses.

If workers are good at anything, they are good at excuses. Part of the training for the job, it seems, is to have a ready-made list of hundreds and hundreds of excuses for why things are done a certain way or why they can't be done.

Discount every excuse you hear. Don't believe excuses. Insist on getting what you want because you can—and you should.

"It's wonderful listening to these guys figure out quick reasons why they did something. Basically they screwed up," Penny Moser says. "And they hope to god you're dumb enough to accept their excuse. I imagine that most people probably are."

Excuses fall into two categories. Category I excuses explain why something can't be done to your house. Category II excuses explain absences, tardiness, and altered states of mind. Category I excuses can usually be overcome by firmly insisting that the work be done the way you want. The success of these excuses is predicated on your ignorance. Category II excuses often are indicative of problems to come. Workers with "family problems," for example, are telling you that they probably would rather be elsewhere and that their minds certainly are elsewhere.

CATEGORY I SAMPLE EXCUSES

"We have to let this base coat of paint dry, so we'll be back on Thursday." (The worker is probably running three jobs simultaneously; almost certainly he is if it's a latex base coat that's already dried by the time he utters these words.)

In one case, an electrician said he couldn't wire an appliance because the instructions weren't there. Paul Locher found the instructions right there and figured out that the problem was the electrician couldn't read. (This became evident when the electrician asked Paul a day later to help him fill out a job application.)

"We can't build that here because the tree branches are overhanging where the wall is going to be." (There are saws.)

"I can't dig a hole here because there's asphalt (or concrete) in the way." (Hand them a sledgehammer, pick, and chisels.)

"It won't fit." (Saws, razors, knives, axes, and other sharp implements are designed to counteract this excuse—and no, you don't use the tools on the subcontractor.)

"You built the footer wrong, and I can't lay block on it." (Poured cement can create a new footing.)

"Can't get the equipment into the house." (Have them build a ramp.)

"We don't have a ladder that will reach that." (Build a ladder out of 2 × 4s, or rent scaffolding.)

"The guy who designed it doesn't know zip about installation." (Redesign it.)

"The salesman doesn't know what he's talking about, and we can't do it that way." (Call the salesman and get him to arrange to do it, or renegotiate the price.)

CATEGORY II SAMPLE EXCUSES

"My truck broke down." (The Number one excuse.)

"My wife is going to leave me if you can't advance me $500."

"I won't be in for a week. I have to attend my cousin's funeral."

The painter couldn't come back to work because he was going to California for the Super Bowl. (In this real case, he never came back—you can imagine how he was able to afford the trip and tickets.)

"I can't do the job unless you pay me in cash."

"Three of my employees just quit."

Workers can be divided into three groups: Those who solve problems, those who make problems, and those who don't want to do any work at all. The first ones are great to have and will work with you, give you options, listen to your wants, do the work you want, tell you the least expensive option, and get the work done. The second type, those who make problems, require more patience, oversight, and firmness on your part. You may have to think like the workers to motivate them.

Those subs who don't want to do the job at all will complain a lot but can eventually be coaxed into doing the work. What are the warning signs of a contractor who doesn't want to be there?

"We don't have the right tools."

"They didn't deliver all the materials."

"The mason's not finished, and I can't start before the mason's done."

"The materials won't be delivered till late afternoon."

"We have to wait for our boss to come and show us what to do."

Your job as homeowner is to tell them, "I'm really sorry all your materials aren't here, but as long as you're here you might as well as get started." Smile and say, "Let's see how far we can get!"

What Hunting Season Has to Do with Your Renovation

Home renovation is governed by your local hunting and fishing regulations. You may wonder what home renovation has to do with fishing and hunting season. Workers, especially carpenters, will never but never work on opening day of hunting or fishing season. They'll work on Christmas or Thanksgiving— no problem—but never on opening day of hunting or fishing season.

In California, the words are different but the tune is the same. Tom Boyd of Studio City, California, reported that the men he hired to hang drywall didn't come to work because "the surf was good that day."* Mother's Day, Labor Day, and Christmas are days when most workers are available, however.

Don't forget bow season and the first day of the races too. Bow season in some states lasts for weeks.

Sometime in July 1994 the World Cup soccer match will be held in the United States. During that month, all renovation will stop, or at least the workers whose nationalities are represented by the World Cup teams won't be showing up for work. If the workers have been watching the soccer match from a bar before they turn up at your house, don't expect any productive work from them for the next few days either.

Many workers like to drink. There's an old joke, Do painters drink because they paint, or do they paint because they drink? There's no definite answer, but you can expect that some tradesmen who are working on your house are not going to be hang-

*Barbara Gray, "Long Haul: Three Homeowners Say Trouble By Yard Yields Profit by Mile,"*Los Angeles Times*, (July 28, 1988): part 5, p. 12 column 2.

overproof. It goes without saying that there should be no drinking or drugs on the job. Construction is dangerous to start with; once alcohol or drugs are introduced, the chances are increased that a worker might overlook something and fall through a floor, or cut himself, or have some other accident. The stories abound: High roofers fall off roofs, drunk painters lean ladders against glass, a concrete contractor tries to pour a driveway on a 10-degree slope running the wrong way.

What a worker does on his personal time is nobody's business, as long as he is fully functioning on the job.

Who Are These People Who Tear Apart Your House?

Other generalizations apply to contractors, no matter who they are. While these generalizations may appear to be a commentary on the workers, they actually are more a reflection of your assumptions about how contractors should behave, versus how they are conditioned to behave.

Many (in some areas, most) contractors and subcontractors book more work than they can reasonably accomplish in the alloted time. If you realize and remember this, you can coax workers into making your job the first one they complete.

Never assume a subcontractor can or will read your notes. Illiteracy ranks high among workers; others won't find your notes or will read them and then instantly forget. (When you do use notes, bear in mind that *notes* is actually the wrong word. Pictorial signs—seriously—at least 8½ × 11 inches are required to get the message across.)

Never assume that English will be a common language to everyone who works in your house. And never assume that the contractor will speak the same language as his subcontractors.

Never assume that everyone who works in your house is honest and uninterested in taking your personal possessions.

Never assume that a worker will stop his work where you want the work stopped. Your instructions on which rooms to paint may be clear to you, but not to the painter, who sees no reason not to coat everything with latex white.

Always assume that workers want to get their work done as

quickly as possible. I know this sounds like a falsehood; probably you've been waiting weeks for the electrician or plumber to return as promised, but I promise you it's not a lie. You're probably thinking, Work quickly!??? These guys *never* show up when they're supposed to! But there's a difference between getting to your house promptly (something not often done and the reason for the assumption that workers work slowly) and actually working swiftly. You may have a hard time coaxing a plumber into your home as scheduled, but once he's there he will want to work as rapidly as he can, no matter what's in his way. If this work involves walking over newly laid tiles, so be it. If this work involves your not being able to use the toilet for a few days, that's the way it has to be.

Don't mix union and nonunion workers. In some renovations, this won't be a problem, but at other times mixing union and nonunion workers is like having antiabortionists and Pro Choicers stuck in an elevator together for a day. Fireworks is the only possible consequence. One contractor (who preferred to remain anonymous) was hired to bring his heating and cooling team to put in a HVAC system. "We were a nonunion shop," the contractor said. "We went in and started work. It was a three-day job. The first day, everyone was pals, saying 'Hi, how ya doing, where ya from?' The next day they asked us about our union, and we gave our street address as our local's address. That got us through day two. On day three they came out with pistols and told us not to show up again."

Never assume that the part the subcontractor is about to install is the part you ordered. Check the newly arrived wood against your specifications; check the plants the landscaper is about to put in the ground against the pictures in the book.

Never assume that the plethora of subcontractors have places to store materials. So plan on designating a storage place for refrigerators, molding, and sinks before the parts arrive.

Check or test—or have an expert check—all vital components.

Never assume that the people who work in your house ever will have heard of Emily Post or Miss Manners. They'll probably be slobs. "Slobs," you say. "What's the big deal? After all, with all that dust around what difference does it make if the workers are messy?" Here's the difference it makes: An electrician with grease-coated fingers puts in an electrical outlet and leaves perfect FBI fingerprints on your newly painted wall. The painters, who are working late, have a cookout on their own hot plate and leave food at the campsite, which spontaneously generates fruit flies (this happened to us). Be firm to the point of insisting that workers wash their hands before touching newly painted walls.

Recheck by calling to make sure that the workers are going to come. Some tradespeople will keep their appointments—with the last person who called them. Since you don't know who else wants a particular worker, only by continually calling will you be able to get them to appear.

In the old days, one tradesman might perform a multitude of tasks—carpentry, plumbing, electrical. The tradesman (the term *subcontractor* didn't exist in previous eras) took the work personally. He felt a part of the project, almost as if there were a part of himself in the house. This was back in time when physicians made house calls and knew their patients by name, not by their chart.

Specialization means that each of the trades cares about his particular work but not the whole project. It's rare to find someone who takes a personal interest in the house he's working on. Everything is fast, specialized, standardized. This breeds indifference to the whole house. This attitude doesn't preclude quality work, but there's a difference between good work and concern for the house or apartment as an entity. When you find a tradesman who does feel a personal involvement—almost a relationship—with the house, you'll know what I mean.

Among the most common phrases that homeowners hear from workers after telling them what to do is, "You don't need that" or a variation on the theme, "You don't have to do it that way." Workers often tell general contractors this kind of thing, because after all, you never know what you can get away with. As a rule, workmen are not qualified to make such assessments. What they really are looking for is a mechanism to make the work easier for them, to your detriment. The only way you can answer "You don't need that" is to know the way the pieces of your renovation puzzle are supposed to fit together. This comes from a basic understanding of what each worker is supposed to do and what he's trying to do. Let's say you're trying to span a load-bearing opening such as a doorway arch and you want to make the opening ten feet wide. When you get into openings of that size, you should insert a steel plate into the *header* (a double piece of lumber that supports the joist horizontally). This header with the steel is called a *flitch beam*. Ordinary carpenters do not like to deal with steel and will tell you that the steel could be replaced with a piece of plywood (wood *is* their medium). You can either say, "Do it with the steel" or, what most people say, "Do what you think is best." But "best" in this case is likely to be what's best for your carpenter, not for the wall. A smart approach would have been to order the materials you require and have them on the job prior to the contractor starting work. This forces him to incorporate the materials into your design. His standard comment will be, "You just wasted money," but you can't put a price on the reliability of load-bearing walls. Anyway, if the material's there, he's resigned to using it. This kind of problem can be averted by having sound, detailed plans and checking that your contractor sticks to the plan.

You can find these examples of laziness with almost any contractor. A mason may tell you you don't need to fill the cells (the openings) of the cinder block with cement. He doesn't want to spend the money or time to fill the cinder block. He gets paid the same for the job, whether the cinder block is filled or not. He'll tell you the blocks don't need cement if he doesn't have cement on site. Who wants to wait for the right materials? Just do the job with what's at hand—that's rapidly becoming the American way on renovation sites.

Workers will always try to tell you that you don't need certain work done, if they'd rather get ready to go hunting. Always. Or they will simply skip doing it without bothering to mention the subject if it looks like you won't notice. A carpenter will tell you that you don't need a termite shield or sill insulation; a heating contractor will tell you he can run a duct through a load-bearing wall without a problem; a painter will say that you don't need a special paint for high-humidity areas; the electrician will say that . . . well, you get the idea.

You should ask each contractor to look at the job personally. It's very important that the people who will actually do the work review the scope of the project prior to beginning it. This allows you time to prepare for their needs, making their job easier and keeping costs down. For example, only by visiting your house in advance could a plumber tell you, for example, that you're going to have to take out the cabinet before he starts. Getting the actual workers to visit the site also gives you a chance to review the job's scope for your own benefit; that way, when they start work you've talked and walked it through, and all you have to do is make sure that the performance matches the plan.

It's a worthwhile safety mechanism to have the next subcontractor examine the previous sub's work before it's completed. While you want to avoid setting one subcontractor against another, it's okay to ask for opinions, especially if you aren't certain what to look for. The subcontractor who has to build on the previous sub's work will know what to look for and will give you a list of what's necessary to properly complete the job.

The People

THE DEMOLITION CREW

Is this a fun job or what?

Needless to say, demolition crews can do the most damage to your house in the shortest amount of time—in fact, that's what you've hired them to do.

Almost every renovation job requires a certain amount of demolition unless you're simply building a separate garage. Renovations of older houses need more demolition than newer houses due to the nature of their design and functional obsolescence. Typical reasons for demolition include a wish to open the space; bring in more light; expand kitchens, bathrooms, and other rooms; discover what's behind walls; bring the house in line with modern needs; and replace walls.

Have a very well-thought-out demolition plan prior to starting the work. You need to make sure that the demolition crew understands what a load-bearing wall is. Understands what a load-bearing wall is. Understands what a load-bearing wall is. This is not a typographical error. One way to ensure this is to hire a master carpenter to work side by side with the demo crew, since he understands the structural needs not only of your house but of the project-to-be. If you have an on-site carpenter, you must tell the demo crew that the carpenter is the boss, because usually demolition crews don't like bosses looking over their shoulders. They don't get paid much, so one of the perks of the job is to be able to swing away freely with their sledgehammers.

Who are these people who wander around your house swinging sledgehammers? They can be professional demolition crews. Many of these crews may have been trash haulers who have broadened their careers by expanding from the simple picking up of trash from the ground to the more complex putting trash on the ground and then picking it up. Others may have built those muscles in a prison yard. They are highly efficient at what they do, but you aren't going to find ex-airline pilots working at this. Demolition crews probably make five to seven dollars per hour; the foreman or company owner might make twelve or fifteen dollars an hour, depending on how many truck loads and dumps he makes. A good demolition crew has as its crew leader someone who has actually been in the construction business. Many general contractors have their carpentry crews do the demolition for obvious reasons; control of the job, scheduling, placement of openings, knowledge of the dangers of removing parts of houses, and, of course, extra cash. A home remodeler handling the subcontracting himself will rarely find a carpentry crew that enjoys doing its own demolition.

The tools of the trade for demolishers are

Jackhammers for concrete,

Twenty-four pound sledgehammers for brick and block walls,

Chain saws or sawzalls for stud walls,

Wrenches for pipes, and

Filter masks.

One caution: Make absolutely 100 percent sure that all utilities are turned off in the area of demolition. Power lines will short and water and gas pipes will leak if they are not turned off. Flip the master switches (at the street if you must).

The best demolition crews will remove items from the house part by part with special attempts made to save moldings and antique bathtubs, windows, doors, and trim. But don't count on this. If there's something you want saved that might be considered trash in any way (an antique floor register, for example), save it yourself. Or guard it until it is possible to remove the object. A lot of workmen will come in and just start swinging the sledgehammer at everything—sinks, bathtubs, toilets, light fixtures, marble floors. For them, the smaller the pieces, the easier it is to fill up a trash can.

Spend time with your demolition crew and make sure that they remove things on a piecemeal basis. If things are taken apart piece by piece, you get to see what's behind the wall, a real plus for your project. You can modify the old systems for future use without having to rebuild. For example, if there's an air duct behind the wall, you may not need to put in the new duct you were planning.

Demolition produces surprises. You don't know what's in or behind a wall until you smash through it. Be prepared to modify your plans based on what you discover. This is another reason to have the builder or carpenter on the job during demolition.

Observe from a distance. When the demo people swing a twenty-four-pound sledgehammer at a wall, and all you want is the plaster taken off that wall . . . well, don't be standing on the opposite side of the wall when the hammer hits. Wear appropriate protective clothing. If you suspect asbestos, wear a mask or asbestos suit. Same thing if lead paint is being stripped. Don't worry if you look like a Martian. At least you'll be a healthy Martian.

Demolition is not a dainty activity.

THE CONCRETE POURER

For building a house or an addition, or for building a footing or pouring a foundation wall, the concrete pourer is one of the first people you'll encounter.

Concrete is the stuff that all houses sit on. It doesn't matter what the design of your house is or how little footing there is, concrete is there. The various specifications for concrete (called *specing* by people in the business) affect the remainder of your project—everything the concrete stands on. You need to have a certain pound-per-square-inch *(psi)* strength with the concrete. What follows may sound like math, but I promise it isn't. Upon the few numbers that follow rests your entire addition or house, so bear with me for a moment. Concrete rated three thousand psi—which you can buy at the local concrete shop—is called *six bag.* Although most jobs require only two thousand psi (four bag) concrete, it's better to take a higher-rated concrete because it provides a stronger foundation. For your own sake you want at least twenty-five hundred psi concrete (five bag); six bag is better still.

Phew, not too complicated so far. We're almost done with the numbers.

The second spec you should be concerned with for concrete is called *slump,* a measure of how fluid the concrete is. *Four slump* enables concrete to be molded in any fashion that a job may require, due to the concrete's lack of water. The most useful type of concrete is in the five- to six-slump range, which allows a certain fluidity and ease of transfer from the truck to the site. Use anything over six slump, and the materials in the concrete— gravel, sand, lime—separate and want to float into different layers, giving the concrete a tendency to crack. This is bad for your house.

If you know something about slump, however, you are likely to end up in an argument with the concrete pourer because he wants the easiest job, which means using a high slump, such as an eight slump. Eight slump flows more easily from the truck but won't hold up as well as six. With eight slump, you don't have the colloidal suspension you need for a secure foundation.

Eight slump is bad. Six slump is better.

The slump number is determined by the amount of water that has been added to the dry mixture. Water is added at the plant, so the pourer probably has the correct slump concrete in his truck. What pourers like to do is add more water on site to make the concrete easier to move. Never let a concrete pourer add water to the concrete on the job without adequate supervision! The concrete will set up while sitting in the truck, so some water may be needed to get it rolling, but only a little water. How do you tell? You don't. You have an expert on site.

Better still, have an engineer or builder on site when the concrete is being laid. Remember, when the concrete is being poured, the foundation to your house or addition is being laid.

At this time it's valuable—maybe crucial—to have an engineer examine the concrete. The concrete pourer won't care about your house after the concrete's been poured; he probably won't even remember your house. Even if there are two pourers watching over the event, they're probably friends and aren't likely to want the other to go back to the shop to get more material. Getting more concrete is a real hassle for pourers: Sending the truck back ruins their schedule for the whole day. On the other hand, if the pourer knows there's an engineer or builder on site, he's more likely to do it right—if the builder insists. A cracked *footer* (a footer is what supports the foundation) is one of the most costly after-renovation repairs you can think of. (Cracked footers also come from other causes, so it may be difficult to blame the concrete pourer later. Noncompacted soil or poor weight distribution can also contribute to a cracked footer.) Get the first step of your renovation right.

There's a curing time for concrete. It usually takes about thirty days (depending on the temperature and humidity) for concrete to reach to 97 percent strength. For the first twenty-four hours, you can practically cut the concrete with a butter knife. Resist the temptation to use your new footer as a makeshift shuffleboard court. During this time, don't put any load such as foundation walls, ladders, trucks, or supplies on the floor. Wait a minimum of forty-eight hours—and seventy-two is better. If that sounds like a long time, it's nothing compared to how long it will take your new furniture to arrive. Be patient.

Concrete is strong stuff, but you can always make concrete even stronger. Using steel rods *(rebar)* is the best way to improve the strength of the concrete. If you are in any measure unsure of the future strength of the concrete, add rebar or wire mesh.

I can't stress how important it is to have someone watch over the concrete. Foundations do crack. Don't let yours be one of them.

Paul Locher once was called in to repair damage in a project in which the homeowner wanted to save money by being his own general contractor. Because the homeowner didn't know what to watch out for, the concrete pourer took advantage of him in almost every conceivable way: The pourer did not dig the footing (the poured concrete) below the frost line; the concrete was poured in below-freezing temperatures on frozen ground; and the slump was too high. Also, no rebar was used, and the concrete was not insulated to maintain its temperature during the cold season. As concrete cures, it produces heat and pushes the water out; insulating the concrete would have kept it warm and prevented ice crystals from forming while it still contained moisture.

This house had star fractures in the footing. The concrete had expanded as it froze, then as it cured it began to shrink. Alas, because of its size, it couldn't shrink evenly.

The footing had to be removed. Fortunately, the cracks appeared before the foundation wall was put on and the foundation wall was delayed. If this had not happened, the homeowner might have had to redo the foundation after the rest of the house had been built.

STONEMASONS

Skilled stonemasons are expensive and often hard to find. But even a skilled stonemason doesn't always do the job right, as Robert and Mary Millstone of Rockville, Maryland, discovered. Masons are not engineers and they are not designers, though they may act as if they were. The Millstones wanted to add a fireplace to their living room but found that their mason didn't pour the chimney's foundation deep enough or tie it correctly to their fireplace. During the winter, the force of frozen water that had seeped under the house detached the chimney from the house. Mary Millstone says, "The first man could have done it, but he was lazy and we didn't know what to look for."* What the Millstones learned from this fiasco was: Hire a professional to oversee the work when the particular task is technical or could cause disaster when done incorrectly.

At a minimum, know what to look for yourself.

The surface of walls should be uniform, especially on the side of the house that confronts the worst weather. There ought to be no cracks in the cement at this stage (later, a staircase pattern of cracks through the brick indicates a shifting load.)

If your house has a *veneer wall* (a wall covering wood framing)

*Ellyn Bache, "Home Remodeling . . . Almost Hassle-Free," *Washington Post*, (November 21, 1981): E25.

holes should be left in the brick wall to permit a balanced air gradient on both sides. Even brick walls must breathe.

HEAVY EQUIPMENT DRIVERS

Don't assume that just because the work is happening outside your house nothing really bad can go wrong. Remember, outside is where all the stuff (electric lines, water pipes, gas lines, sewers) waits peacefully for some screwup.

A Delaware builder was regrading around a house. He got a contractor with a machine to come redo the grading and dig a footing for a retaining wall. It was relatively minor work utilizing a 450 loader (a type of bulldozer). The builder says, "The day of the work I met the driver of the machine for the first time and proceeded to walk the job with him. Standard operating procedure is to walk through the job with the contractor prior to the work commencing. I identified two potential problems: There was a water line supplying the house, which he was to avoid, marked with stakes, and a soft spot in the ground, which he was to avoid to keep the 450 from getting stuck. I also showed him all the other utilities for the house just for his information. I watched him working for the first one-half hour; it looked like work was progressing smoothly." The builder took a break to make a call from a nearby pay phone. The call took five minutes. He continues, "On returning to the site, I discovered that the machine operator had cut the water line, run into the house, run into a telephone pole, and become bogged down in the soft spot, which was now full of water, since he had cut the water line. It was a builder's worst nightmare. There wasn't a shutoff valve for the water on site. It took four hours to locate the valve in the street a quarter mile away. Four hours of water going through a one-inch line at sixty pounds per square inch is about 100,000 gallons, all of which ended up on the lot. A crane had to lift the bulldozer out of the site the next day."

How did this happen? The driver hit the house and then the telephone pole because he was trying to escape the water.

Eventually, the builder had to repair the water line and put a new valve in. To effect repairs on the house's outside walls and elsewhere, he had to jack up the house. He had to replace a telephone pole. There was mud in the yard for three months. Not including lost construction time, this incident cost about $15,000, all because of five unsupervised minutes.

But there's more cost! The excavating contractor sent a bill for the machine time on the job and the cost of the crane to remove the machine. Naturally, it wasn't paid.

What's the moral to this story? Any time you have a combination of heavy equipment, unskilled labor, and lack of supervision, change the rules. Supervise. Period.

Oh, and here's the amusing part: The driver didn't even tell the builder that he had hit the house or run into the telephone pole, hoping that his boss might miss these problems.

In another case, a contractor was installing a new one-inch water pipe to supply a house. This required trenching and tapping into the water main. The backhoe, which was on the opposite end of the bulldozer, was working smoothly, and again the contractor took a five-minute break. On returning to the job, the contractor witnessed a sixty-foot plume of water rising out of the street. Because the temperature was 20 degrees Fahrenheit, the water froze when it hit the ground. The object that was immediately beneath the ice was the backhoe. The driver, who had knocked a tap off the main water line, couldn't stay in the vicinity to do any more work due to the freezing water which was quickly encasing his machine, his pile of dirt, and the road. It took about four hours to turn the water main off, while the neighborhood turned into an ice-skating rink. Everything was under eight inches of ice. Sixty-year-old oaks no longer stood upright; they sagged to touch the ground. Because the temperature stayed in the 20-degree range for the next few days, the crew didn't get much work done. Only after several days of 60-degree weather did the area thaw out. Fortunately, workmen were able to chisel through the ice, cap the broken tap, and restore water to the neighborhood. However, this did not help the builder in his quest to get water to the house.

What's the moral to this story? If you see major work going on in your neighbor's yard, check your insurance policy and close your windows.

INSULATORS

These guys are going to make your house either the most enjoyable and comfortable place to reside or the worst imaginable.

Insulation is made to fill empty spaces where air can pass through unobstructed. Insulation laborers get paid, generally, by the square foot; therefore the more square footage they do in a given day, the more they get paid. The insulation subcontractor will bill by the square footage covered rather than by the square footage used on the job, therefore, there is no automatic checking mechanism to verify that all the insulation that was sent to the job was actually installed. One contractor witnessed an attic inspection in which there was one foot of space between every *batt* (roll) of insulation. That house was not insulated. Skimping on insulation can happen to anyone who does not check the insulation job before drywalling. Look for two elements after the insulator claims to be done: First, there should be no visible gaps between pieces of insulation; second, no exterior construction material should be exposed from the inside.

PLUMBERS

As with most workers, generally the person you talk to about doing work isn't the person who will do the work, unless you hire a sole proprietor. If you hire a company, a salesperson usually will come and talk with you about the job, but the salesperson may be so far removed from the work that he really won't have any idea of the problems of your particular house.

Some general contractors have a preference for sole proprietors. They have a favorable cost to experience ratio. With a sole proprietor you generally have more flexibility in negotiating prices, dividing work, and supplying outside labor (for example, having a laborer dig a trench at seven dollars per hour rather than leaving it to the plumber at thirty dollars an hour). You can always find some way to negotiate on price with a sole proprietor.

Other contractors prefer big plumbing companies. Companies have better liability insurance and training programs, possibly greater reliability, and more fine-tuned service. They are often faster workers: They like to come in, get the job done, and exit. This can be good or bad. Large plumbing companies need a large volume to cover their overhead.

The smaller guys don't need the volume but may skimp on things like workmen's compensation insurance, liability insurance, and paying for their materials promptly (which can affect you later). But sole proprietors may take more interest in your particular job; they're working for you rather than for a company. Repeat business depends on the quality of their work, not their advertising. Sole practitioners are not paid by the hour, but at a negotiated fixed price.

Plumbers like to do things their way, which is not necessarily the way you want them to do things. Bob Adriance's plumber didn't want to follow the plans drawn by architect Robin Roberts. "All the mechanical equipment, the wires, the pipes were supposed to be recessed," Bob says. The plumber ran a pipe down the middle of the guest room. I had to argue with him for quite a while to get him to move it. You could tell he hadn't looked at the drawings; he just put the pipe wherever the hell he wanted."

Much of the time there is only one reason a plumber doesn't follow the plan: He doesn't feel like it. Your general contractor—or you—should make sure that he does feel like it.

Many workers, plumbers included, don't look at plans. They rely on the contractor to tell them what's in the plan. When a worker reads a plan, chances are he will not follow it correctly.

Plumbers have to connect large pipes in small spaces to immovable objects. As a result, they will cut whatever is in their way, including structural elements. Plumbers are notorious for cutting large holes in structural beams. Architect Ted Fleming says, "I've seen plumbers who have no concept of gravity. It's amazing that some of these old houses are still standing when the plumbers are through."

Here's a short list of how plumbers can become bummers:

Reversing hot and cold lines;

Not venting drain lines;

No traps in drains;

Cutting into load-bearing walls;

Poor solder in copper pipes, allowing leaks in pressure lines;

Poor gluing of drain lines made of PVC;

Poor gasket placement in cast iron; and

Putting pipes in rooms instead of in walls.

Anytime you have different materials joining together—new PVC drain lines connected to old cast-iron piping, old galvanized water lines leading to copper—there is a danger of leakage. Now and later. This general principle—the different materials connection principle—is as true for plumbing as it is for any other aspect of home renovation.

Water damage isn't the only potential hazard you face. Plumbers install gas lines too. Gas line connections are all threaded with valves, and they all have a tendency to leak. Any system under pressure, such as your gas lines and water lines, must be pressure-tested, no matter how good your plumber is, what company he works for, or how long he's been in business. You *must* pressure-test the lines. For copper water lines, a regular screw-on pressure gauge does the trick. For gas lines, the best is a Coleman gauge. A Coleman gauge will tell you if your pressure drops even by one one-thousandth psi.

Plumbers are required to pressure-test all new construction but have a tendency not to pressure-test on renovations. "Phew! The pipes are connected, the systems work, as far as the homeowner is concerned my job is done . . . It's Miller time!" How do you make sure the plumber is pressure-testing? You watch. Say, "I want to see a pressure test." This is extra work for them. They'll tell you it's silly, but what the heck. Be silly. They all know how to do it; they all think it's a waste of time. But a pressure test will catch the one of every one hundred pipes that leak before the leak damages your health, your house, or your bank account.

What do you see when the plumber does a pressure test? A pressure test uses air to test the system, not the material—water or gas—that normally flows through. If air bleeds out of the pipes, indicating a leak, there's a drop in the pressure gauge. Any pressure drop signifies a failing grade for that pipe system. The pipe should hold the pressure for half an hour to twenty-four hours (check local plumbing regulations for the standards in your community). Call the plumbing inspector's office or the gas inspection office for the exact measurement. These offices may even help you do a test.

Plumbers like to test lines by using the materials that are normally pumped through the system as his testing agent. That is, he turns on the water and looks for sprays, or he turns on the gas and holds a butane lighter to the joints in the gasline. This is not safe for you. Sometimes a plumber will test by putting liquid soap on the joints and seeing if bubbles form. These alternative tests require analysis by the person doing the test; it's not like reading a gauge. If the gauge shows that there's a leak, then one of these alternative tests may be used to try to find out where the leak has occurred. You know the analogous situation: Your car tire leaks, and you hold it under water to find the source.

Usually plumbers skip the pressure test and just check the system by using these visual methods. That's not satisfactory, because the readings are not accurate.

Never let the plumber test gas lines with lighters. Unless you like pyrotechnics.

For gas lines only: A basic rule in plumbing is that valves leak; about 50 percent of all valves have some leakage. Your plumber may want to remove the valves before doing a pressure test. Is this okay? Probably. Passing a pressure test without any valves tells you that the main joints are okay, and that's what's most important. Valve leaks are generally minute.

LABORERS

These are the guys who think nothing of putting sopping-wet sweatshirts on top of your original Calder sculpture. However, the laborers are the people who can make your renovation successful and enjoyable if utilized and managed properly.

Laborers are used for

Sweeping,

Trash collecting,

Digging ditches,

Backfilling,

Demolition,

Transportation of material,

Moving furniture,

Landscaping, and

Basically all the grunt work that you don't want to pay the contractor to do.

Unskilled laborers are far, far cheaper by the hour (five to seven dollars an hour and even less in certain parts of the country) than skilled tradespeople.

The major problem with laborers is that they don't understand the scope of the work (or English necessarily) and may be forgetful even if you explain it to them. You need to spend time with these guys if you are going to realize any efficiency from their use.

Your contractor probably has his own laborers. You can make a deal with the contractor to keep the laborers busy. If the laborers finish a particular task, you can get the contractor to tell them to do something else. If the contractor doesn't have laborers, you might consider hiring your own—kids in the neighborhood or people who advertise on local bulletin boards.

Many subcontractors don't like to perform certain tasks, though these tasks may be necessary for their work. This is where laborers are kept busy.

No one likes to use shovels (though digging holes is necessary for the plumber to lay pipes, for the electrician to wire outdoor lights, or for the mason to lay a foundation). Carpenters don't like to do drywall repair work, even though it will help them do a cleaner job. Carpenters also don't like to break apart walls, though this is a prerequisite to constructing new walls. And painters don't like to move fragile objects out of their way.

When a laborer does something that a skilled tradesman might do, it's best to keep an eye on that laborer—for your house's sake.

CABINETMAKERS

This person may be your carpenter, or he may not be. There are plenty of reasons why you should let your carpenter build your cabinets: He's on the job; he'll be less expensive than hiring a new person; by now he has a sense of what your tastes are; he knows the dimensions of the space where the cabinets are going to go; and he's probably going to be able to work swiftly.

And then there are some reasons why you may not want to employ your carpenter to make your cabinets. Despite his amazing array of skills, many carpenters are not as good at building cabinets as cabinetmakers are. Simply put, people who make cabinets for a living are going to be good at it. (A generalization, I know; so don't hold me to this statement.) By way of analogy, when I was at Georgetown University Hospital for surgery, the medical students usually put in my IVs. It hurt. When the registered nurses put in the IVs, it barely hurt. Why should nurses be better at something than doctors-to-be? Because medical students have only done a dozen or so IVs, while nurses have done thousands.

Every day you'll look at and use your kitchen, bathroom, living room, and basement cabinets, so they might as well shine. But be sure to examine the credentials of your cabinetmaker. There's only one way to do this: Visually inspect cabinets he has made for others. "I didn't check out the cabinetry before it was made," laments Katharine McKenna, who had her New York City apartment renovated in 1989. "The cabinets were cheap, and they look cheap. I took the guy's word for it when he said he was 'the best.' I feel I got ripped off. If I'd known the cabinets were going to be so bad, I would have shelled out more for better. I hired this guy because the contractor recommended him."

Throughout your renovation, your contractor will recommend various subcontractors. Some, like plumbers and electricians, may have only an indirect impact on the way your house looks. Others, like the cabinetmaker, painter, and tile layer, may have a large influence on the visual impression your house or apartment makes. When this is the case, it pays to take a look at the recommended subcontractor's previous work or portfolio.

PAINTERS

Every time I walk out of my house and see the tree on our front lawn that the painters used to clean their brushes, I want to commit a crime. I try hard not to see that tree. It doesn't do any good to go out the back, because they coated two trees in the backyard with white oil-based paint. Actually, at nighttime I don't get too angry, but only if there's no moon to illuminate the trees.

One of the first instructions you should give painters is where they should clean their brushes and their hands. Designate one sink, preferably a basement sink, and insist that they use only that sink. Make it as clear as you can that all other faucets are off limits. Physically disconnect those sinks (or put tape over them.) Unless instructed otherwise, painters will use any convenient sink. You might also consider telling your painters that they must rinse the designated sink thoroughly each time after using it, although I wouldn't count on that happening.

In the same breath, tell your painters not to clean their brushes outside. You might assume that telling them to use only one sink is the same thing as telling them not to clean their brushes outside. It is the same thing, but not to painters. Again, be very clear: Dry paint clings like fleas to a dog.

You may have noticed that the painters brought with them drop cloths. Useful items, these drop cloths, but only when they use them.

Generally, painting follows a particular order: Caulk, spackle, patch, sand. If rollers are used, corners and edges must be painted first with a brush. This order ought not to be reversed.

Painters have to move ladders, paint cans, lamps, brushes, and those wooden stirring sticks around the house all day long. The last thing they want to do is keep moving drop cloths from place to place. Painters have three ways of rationalizing their failure to use drop cloths:

They are professional painters and won't splatter much paint anyway,

Nobody notices a little splattered paint, and

Those few drops of splattered paint can be removed with a straight-edge razor.

A razor? On your wood-panelled cabinets? On your marble bathroom sink? Remember the discussion in Chapter 6 about specialists? Painters are not specialists at removing paint because their job is to put it on. The only person I would trust to remove paint with a razor is a mohel.

Don't let painters splatter paint around; require them to use drop cloths. Everywhere. If they say they can't get a drop cloth to cover something—usually a vertical object like a refrigerator—tell them, "I know you can."

Painters are slobs. This is an overgeneralization, for certainly Cezanne and Picasso were relatively neat. Take for example what happened in this renovation of a large, old house on St. Charles Avenue in New Orleans, a story related to me by architect Ted Fleming. The owners were doing an extensive renovation, so they convinced the carpenter, who also acted as the de facto general contractor, to move in. This carpenter was exquisitely talented. He cut wood by hand and the lines were perfectly plumb. After the attic was finished, the painters were let loose.

Reckless is the best description of them. The painters brought up their tools and a half-dozen six-packs. The carpenter told them to clean up their mess. And they did. They found the carpenter's laundry hamper and used his clothes to clean up the paint. (There are a couple of other lessons to this story, explored elsewhere in the book, including that tradesmen are not above sabotage if you anger them. Treat tradesmen as your would treat a buzzing bee.)

Sometimes painters use sprayers. If you are going to spray paint* your house, don't let them water down the paint, as they often like to do. Watered-down paint will go through the nozzle that they have on hand more easily. Instead, they should get a larger nozzle for higher-viscosity paints.

Occasionally you'll encounter dishonest painters, painters who pour store-brand paint into name-brand containers. If most of the painter's cans aren't sealed when he starts the job, suspect a crook.

THE ROOFER

This is an important person. He helps keep the rain out. Rain on the inside is bad for houses.

If the roofer—or anyone else—is installing a skylight, make sure that it's done right. That is, ask the roofer directly, "Are you sure the skylight is installed correctly?" All skylights eventually leak. Will leak, not *might.* A good installation makes the difference between leakage in six months or in six years. Skylights need to be built up, flashed, and counterflashed so that water doesn't accumulate under the skylight and leak into your house. If you're the neurotic (or sensible) type, it's worth having someone else take a look to ensure that the skylight was installed well.

One general contractor said, "All roofers are crazy." Think about it. It's scary up there.

Roofers, like everyone else who work on renovations, are prone to mistakes and laziness, either of which can be catastrophic. If the roofer doesn't hammer the nails all the way down on an asphalt shingle roof, whenever anyone goes up on the roof

---

*In the first draft of this book, I typed "stray paint." A Freudian slip, I'm sure, but a slip that isn't too far from the truth.

or whenever there's weight on the roof like snow, the nails will pop through the shingles. Roofers don't always get the nails in all the way, so if you have occasion to get on the roof, take a look at the nails. Better still, hire an inspector.

On the subject of nails, there's a proper nail for every task. Often the right nail is codified in your local building regulations. For example, galvanized (rustproof) nails must be used on roofs (as well as decks and other outdoor structures). Galvanized nails are more expensive and require an extra trip to the supply store, so sometimes a roofer may skip this step, or use some galvanized nails and some nongalvanized ones. If he does, the nails on the roof are guaranteed to rust, probably in as few as three years. From that moment on, you have a weakened roof.

Also, make sure that all roof areas, the attic, and eave spaces are well vented. If moisture gets trapped in those spaces, it will rain inside your house.

THE EXCAVATOR

This is the person who digs a big hole near your house. The hole obstensibly is for an addition, but soon after it has been dug you'll discover that it doubles as a manmade lake. Excavators usually don't work in isolation; that is, they are hired by a general contractor, whose task it is to see that the hole is filled with something useful.

"When we would come home from work," Michele Sands says, "we would see piles of dirt. We were thrilled. A year of planning was actually happening." The addition that Michele and her husband were building was to be magnificent: a bedroom on the edge of the woods with a full glass ceiling from which to observe the stars and sky.

"They didn't carry away enough dirt," Michele says. "They didn't calculate how much dirt they needed to carry away. They did an excavation and the next step was for the concrete man to come in here and build a retaining wall. The concrete man looked at the hole and said, 'This does not conform with OSHA [Occupational Safety and Health Administration] regulations, and the hole isn't big enough.' So then the excavator comes back with the machinery and has to do it again. At this point, we were ten weeks into the project and we still didn't have the proper foundation."

Her husband adds, "Their attitude was, 'We really can't plan this because dirt expands when you dig it out.' I don't think anybody went through calculating how much room they would need to store or move the dirt."

Excavators don't always know how deep to dig the hole and don't make allowances for storing or moving dirt. Unless someone tells them to keep plowing away with their Bobcats, backhoes, and bulldozers until they are told to stop, they will stop when they feel they're done.

Then it's not only possible, but likely, that you will be left with a big hole for a long while, as the contractor tries to get the machinery back. Or the hole will stay until the contractor is ready to start stage two of the project, building.

FRAMING CARPENTERS AND REGULAR CARPENTERS

When you are constructing a house, you have a foundation laid, and your framing carpenter comes in and gives you the skeleton of the house. The framing carpenters have the house completely—they are the only ones working on the house. Their job is to give you square, plumb, and level construction. As they build, they are to check at each floor level. They check the outside walls, and use *springboards* to move the walls. (A springboard is a piece of lumber that's attached to the floor and the wall on an angle; tension is put on that board to pull the center of the wall into a straight line. The easiest way to check a straight line is to pull a string from one wall to another.) Once the wall is straight, the springboard is secured in place, and the carpenters start the next level of framing. They shouldn't start on the next floor until the first has been checked. That way you know you are getting a square house. The springboards are a temporary brace, removed after the house is sheathed and completely framed. Cross bracing is also required and can be done with wood or metal. At every corner of the house, there's a cross brace that keeps the house from torquing out under stress. (Cross braces run from the floor to the ceiling at an angle.) It's very easy for construction to be done without checking everything that needs to be checked, and easy to forget about bracing, whether it be temporary or permanent. The consequences of not bracing your house can become very vivid. At one construction project in Potomac, Maryland, the

carpenters hadn't put in the cross bracing. They figured their work was complete, and they there were ready to hand the house over for the next stage of construction, which was the brick veneer. However, along came a big wind, just like in the children's story, and figuratively blew the house down. It skewed the house on the foundation; it twisted each floor above the first floor so that the house looked like the start of a corkscrew. This is known by the technical term "a big problem."

But the issue then became, Does the whole thing get torn down and rebuilt, or modified from what's left? The builder's decision was to try to pull the house back into square. What he had to do was place bolts at critical points by putting in I bolts (one-quarter-inch steel) and put in extra plates. To the I bolts he attached three tow trucks, using the tow trucks to winch the house back into square, and hold the house in place while the (new) carpenters went inside and braced the house. (The virtue of resquaring over rebuilding was about $8,000.)

Your framing carpenters should be rechecking every piece of construction, square, plumb, and level. When the carpenter is working on your house, you must have a level and tape measure to inspect his work. Carpenters sometimes think they can "eyeball" straight lines. They cannot. Even the most detailed plans leave room for interpretation. Don't let your carpenter decide willy-nilly what he thinks you want.

Once the skeleton of the house is done, you'll know the inside parts will fit. That is, your sink will slip into place rather than leave a one-inch gap. When the framing is good, everything else glides in, but it's not unusual to find a house that is out of square.

In one particular house that was framed seventeen inches off, the incorrect framing showed up when the cabinets were put in the library. Any time you have custom cabinetry that's square and you try to put it into a space that's not square, it either won't work or will be ugly.

Carpenters are the magicians of renovation. They turn vast open spaces into rooms and other useful places. They turn dull, drab hallways into interesting spaces by adding molding, creating arches, building doors.

Not all carpenters are experts at geometry. The angles they

measure for doorframes and windows are not always perfect. That is, you will notice the mistake.

If your carpenter is in charge of ordering his own materials, verify that he's ordered materials to the contracted specifications. Wood, for example, should be the grade you've selected. Doors are one place where carpenters sometimes switch.

Carpenters are good at finding dry rot. If they're not especially busy, they may notice that the dry rot they've found has extended a great distance in your house, while in fact only a small section is afflicted. Be careful of this scam.

DRYWALLERS

Drywallers can make your house or apartment look great—or awful.

Drywallers skimp in three places. First, the interior drywall may not be as thick as you want. How thick should it be? At least as thick as the walls that were there before and at least as thick as walls in similar houses. Your contract should specify the drywall's thickness. Take a measurement of these other walls and then check yours.

Second, if you've specified that insulating, fireproofing, or soundproofing material be placed in the drywall, make sure that 1) the material is on hand the day the drywaller comes to work, 2) most of it is gone by the end of the day, and 3) none ended up in the dumpster. Naturally, it's best to inspect the drywall as it's being erected.

Third, bathrooms and other wet areas of the house require water-resistant drywall. Regular drywall is gray; water-resistant drywall is green. Cheap drywallers will substitute regular drywall, and a few years later you will have horrible leaks. By then the drywaller will have returned to Portugal.

ELECTRICIANS

If the electricity works, that's all that matters to electricians. Few care if outlet boxes and switches are put in evenly. Most electricians seem to have apprenticed on board oceangoing vessels, where they had to install switches during violent storms.

Why can't electricians put in outlet boxes straight? While this is one of the great mysteries of life, let me suggest that the reason is that electricians don't think you'll care enough to ask them. Another reason may be that electricians don't carry levels.

Perhaps it's a worthwhile idea to loan your level to the electrician and gently say, "This may help you get the switches and outlets level."

But it's the ceiling fixtures that can be a real eyesore when placed unevenly. Inspect the electrician's work carefully, or you may be living in a room whose walls run due north and whose light fixtures point to magnetic north.

If your plans call for down lights, make sure that there is enough room above the ceiling. We wanted fixtures for five-inch down lights—and that's what the electrician put in. But the ceiling space wasn't deep enough for five-inch bulbs, so we had to use three-inch bulbs. It looks funny to have a large socket with small bulbs.

When electricians use plastic outlet boxes instead of metal boxes, they must ground each plug individually, something they generally forget to do. Have your builder check for this.

A house in which the circuit breakers pop every time you start the vacuum cleaner is not a joy to live in. Ask your electrician to run separate wires and breakers for the major appliances such as the refrigerator, dishwasher, and heavy air conditioners. (Many local codes require separate breakers for each kitchen appliance.) Same goes for new areas of the house: Each should have its own breaker. It's easier and less expensive to tap into existing lines now, but you'll suffer later.

HEATING, VENTILATION, AND AIR CONDITIONING (HVAC) WORKERS

Beware of people wielding chainsaws: First rule. They have a tendency not to be careful when they expose a section for work. Electric lines, joists, and load-bearing walls don't have much of a chance against a chainsaw. Unfortunately, a chainsaw is a standard tool for HVAC contractors when they do their rough openings, so watch them carefully. One contractor who was overseeing ductwork said, "I have seen joists cut because the HVAC opening is drawn from above and cut from above, and it isn't until after the fact that you go down below and notice that one of your supporting pieces of lumber isn't supporting any more."

In general the contractor isn't going to notice that an entire piece of supporting lumber has been cut. "Notice," in these instances doesn't mean "see," it means "tell the homeowner."

The workman with the greatest potential to do the most damage is the workman who's lowest on the totem pole—the guy with the saw, the pick, the heavy hammer. He's given the grunt work, and the grunt work is where the worst mistakes are made.

Most HVAC companies are reputable, but dishonesty can be found anywhere in the remodeling business. Some HVAC companies cheat by installing a refurbished unit, pretending that the equipment is new. Inspect the machine's paperwork to verify that it is new and has a full warranty.

The absence of sufficient return for supplies is also another problem. Getting cool air in doesn't help much if the system can't take the hot air away. You should ask the HVAC company to give you shop drawings of the completed system. This forces them to actually calculate how much heating and cooling you'll need rather than making off-the-cuff estimates. Many HVAC companies don't sit down and figure out the supply and return loads, but they should.

CARPET INSTALLERS

He's one of the last people you'll see, since putting down wall-to-wall carpeting is often the final component of a renovation.

Or so I thought. It seems that the carpet installer requires another visit by the painter, or at the very least requires that you take a trip to the hardware store for more paint. The rough bottom of carpet padding feels like tough sandpaper and has the same abrasive capabilities. I've never heard of a carpet installer who, while laying the padding, doesn't scratch the paint off the walls, particularly near the bottom of the walls. (Of course, you can't have the painter in after the carpet installer because you know what the painter would do to the carpet.) You must be prepared to touch up those areas that the installer scrapes.

TILERS

You have to tell the tile layer precisely where you want each tile, otherwise he will determine the pattern. Draw a map. If your tile pattern is complicated, plan to be on the job that day. Other than that, most tilers are very good at their jobs.

> I was home one day with the flu. That day a worker fell through the ceiling in my bedroom.
>
> Bob Adriance, homeowner

10 | Preventing Catastrophe

MISTAKES WERE MADE. So the saying goes.

There's a little more to it than that, however: People make mistakes.

From the contractor's point of view, mistakes "are made." From the homeowner's perspective, there's an individual who is liable for each and every mistake.

In this argument over voice, you are right and the contractor is wrong: Behind every mistake there is an active verb because there is an active person making that mistake.

So much for semantics. What does this little lesson on grammar teach us? Every construction mistake is a preventable mistake—in hindsight.

Some mistakes are forgivable errors, such as having to substitute one model stove for another because your first choice won't fit; or laying the floor tile slightly crooked because the original walls don't make a perfect rectangle; or not noticing that the electrician did not put in an outlet next to the sink until after the room was painted; or finishing the ceiling of a room on the first floor, only to discover that the ceiling has to be reopened to reroute the plumbing in the room above. None of these mistakes is welcome, but none is a disaster, either.

Preventing catastrophe is different from preventing the destruction of your personal possessions. Catastrophe involves mistakes that make your house, or a part of it, come tumbling down.

Real mistakes are of this nature: installing a double tub that, with two people and 150 gallons of water, is too heavy for the floor; ignoring termite damage; removing a water softener in the basement but forgetting to seal the pipe that leads to the outside; installing a skylight without sealing the edges; building without a zoning variance and having to tear the whole addition down. These are—or can suddenly become—unforgivable errors. The destruction of your belongings doesn't necessarily harm your house, but a structural catastrophe damages not only your house but everything around it.

Catastrophes can defy our language's ability to express horror. Take the case of the New Jersey couple who just wanted to expand their house by adding dormers. Toward the end of the project, the house collapsed because the contractor had made it structurally unsound. One of the contractor's tricks was to paint wood black to look like steel beams. Of course there's always worse. Town officials then told the homeowners that if they tried to build their house again the way it was, they would be violating current zoning regulations.

Your homeowner's insurance may not cover these Titanic-grade blunders. It's worthwhile checking and updating your insurance policy to prevent financial catastrophe.

In this chapter I will give straightforward advice on avoiding all sorts of catastrophes and dealing with mistakes as they occur, which in turn, will help stave off large blunders.

When work is done correctly, potential catastrophe still looms over your house. One tradesman can easily—and inadvertantly—damage another tradesman's work. This happens all the time, and the results range from annoying to catastrophic. For example, siding installers damage roof shingles by nailing their scaffolding through the shingles, HVAC men cut through joists, electricians drill through water lines, and drywallers nail through electric lines.

The Right Mental Attitude

What happens inside your head is almost as important as what goes on in your house. First, always keep in mind: Everything can be fixed. There are monetary and psychic costs, naturally, but eventually any mistake can be rectified.

Second, remember that the agony passes. Think back to final exams in college, boot camp exercises, being caught speeding on the freeway. These unpleasant events take on a dreamlike quality soon after they are over; this is even truer with home renovation. When the job is done and you are enjoying your addition or renovated kitchen, the pain will seem part of another world, a world in which you no longer live. When your renovation becomes agonizing, think into the future: How much fun your house will be. The agony will rapidly slide away into a bygone era.

The last—and most important—proposition to keep in the forefront of your mind is that you should be flexible. Be prepared to do things you didn't expect to do at the onset. When the subcontractor who was laying the tile in Michael Weiss's kitchen did a "lousy" job—the glue made the linoleum tile bubble—Michael said to the contractor, "Let's you and I tear this tile up and put down new tile." Laying tile wasn't a job Michael had ever expected to do himself, but because he was willing to be flexible, he has a completed kitchen.

These three tenets will enable you to cope with the disasters that probably will occur.

Knowledge Is Power

The most successful technique for preventing catastrophe is to anticipate disasters. You should look at every element of your renovation as a potential disaster. Ask yourself, What can go wrong here? Examine each step, question every part of the renovation—before the work is done.

Start to gather knowledge about your renovation by getting involved in the design stage of the project. This will help you understand what is to come, especially if you ask why something is in the plans. Why are these steel beams being put here? Why only a thirty-four-inch refrigerator? Why no recessed lights in the bedroom? If you are involved in the plan's creation, you will be able to understand the plans better and know what goes where. You will more easily be able to catch omissions and other mistakes later.

If the doctor said, "Hmm, I don't like the bump on your index finger. Let's chop the finger off," you'd probably have a few questions. The same inquisitive attitude should be directed toward your renovation.

The more you know, the more you will be able to handle disagreements among experts. An engineer might say, "Knock down the interior wall this way." The builder might respond, "If I have to take it down that way, I quit." Honesty has nothing to do with differences among contractors—for many problems there's bound to be more than one solution. What works in one house won't necessarily work in yours. How do you resolve these differences, especially when the wrong solution can spell catastrophe?

The more you know ahead of time, the more you are going to be able to make solutions worthy of Solomon.

Trust Yourself before Trusting Anyone Else

That includes the information in this book. Your house, your subcontractors, your plans are going to be different from every example in this book, so you must be prepared to make your own evaluations. In addition, be prepared to be the recipient of incorrect or incomplete information. Even as venerable a source of information about renovation as the television program "This Old House" is said to make errors. It's been reported that some renovation costs the show gave may have been shy by $200,000 because, unlike a real renovation, most of the materials used were donated by manufacturers. When not in doubt, that's the time to doubt.

When a San Francisco homeowner engaged in extensive renovation noticed that water was running down the inside of a foundation retaining wall, she said that wasn't supposed to happen. A neighbor who had borrowed the homeowner's hose and water hooked up the hose in such a way that water leaked through the wall of the house. Somehow, the homeowner took it upon herself to spray water along the entire side of the house and found that water leaked into the house in several places in the foundation wall. If serendipity hadn't intervened in the form of the neighbor borrowing water, the homeowner never would have caught the problem.

Believe your eyes and instincts.

General contractors want you to pay attention to potential problems. A good G.C. would much prefer that you spot a problem early rather than find it later, when it becomes more costly for the builder to fix.

Supervision Is Worth Every Dollar

Supervision has two immediate consequences. First, when workers know that they are being watched (or that their work is being checked periodically), they tend to do better work. This is human nature. They do better work because, in the presence of company, they feel like giving extra effort and because they don't want to have to do the work over if the supervisor deems it wrong. Second, supervision means that a second pair of trained eyes examines the work being done, and you have a better chance of actually catching mistakes.

I've never conducted a survey, but I would say that the chances of something being done right is two to five times greater with supervised work.

The best thing you can do to prevent catastrophe is to hire a general contractor who is on the job all the time.

Be Prepared to Make Quick Decisions

A well-crafted plan should mean that you won't have to make any decisions as your renovation progresses. But in real houses there is no such creature as a perfect plan. The unexpected always appears. And if you're working on a time-and-materials basis, your entire renovation is based on making daily or weekly decisions.

While making quick decisions isn't necessarily directly related to preventing catastrophes, it has an important effect on your house. If you take several days to decide on an alternative bathroom faucet because your original choice is no longer available, and if you take several days to decide on an alternative window trim, and if you take several days to decide exactly where you want the built-in bookcase to be sited, you are going to disrupt the workers' schedules. Which means that the workers may have to come back at times they prefer not to, or that the workers may have to work more quickly later on, or that the carpenter may have to put in some wiring because the electrician who was at your house before is now elsewhere, or that new work will have to be done around not-yet-completed old work. These events can lead to work of lesser quality and perhaps to mistakes.

Plan for Mistakes

Every worker is going to screw up a little. When dealing with human beings, there's no such thing as perfection. Most mistakes will be minor, but some will leap into the next category, annoying. While it's almost always possible to rip apart bad work and start over, that isn't always the best course. It can be expensive and time-consuming, and the tradesmen might rebel.

Often, fortunately, the tradesman who comes next can rectify the mistakes of the previous worker. A good drywaller can make a badly framed room look nearly square. An alert carpenter can use trim to make the drywalling and the floor work well together. Good painters can fix scratches left in the wall by the electrician. So when you see something that hadn't turned out the way you expected, find out what can be done by the next subcontractor.

Water

TAMPA, Fla., Jan. 4—Leaking water from a toilet may have partially caused an engine to fall off a Northwest Airlines Boeing 727 today.

Washington Post, January 5, 1990

Water, the essence of life, is the enemy of homes. Plumbing is the mechanism that brings clean water in and dirty water out. In pipes, water crisscrosses walls, ceilings, attic spaces, and basements performing crucial and sometimes urgent tasks: washing dishes, running showers, soothing burns, removing stains. But out of pipes, water has even greater powers, and other than fire and earthquakes, nothing can ruin a home faster than a leak.

Gravity and capillary action will compel water from a leaky joint, shower, faucet, or toilet to bisect your house, this time destroying your house in the process. It is for this reason that all aspects of plumbing should be carefully planned and carefully constructed, starting with your choice of plumber.

Water damage can be acute—for instance, a burst pipe that ruins everything in your kitchen. Or the damage can take a long time to occur, such as a ceiling falling in beneath a bathroom. Or, perhaps worst of all, water damage can be slow and creeping. *Dry rot* is the term and the process. It is very dangerous. Dry rot occurs when wood is alternatively wet and dry over months or years; eventually the wood looses its strength and collapses. Just a few drops of water from the tub each time you take a bath is enough to cause dry rot.

Get the best at the best price (you don't want NASA's space shuttle plumber; you can't afford him and he's overrated). It goes without saying that you should use a licensed plumber. You may decide to forgo a plumbing permit here or there, but a permit won't guarantee a good, safe, or complete job anyway. Throughout your renovation you may have been trusting your general contractor's judgment of subcontractors, but when it comes to plumbers you should personally investigate, looking at the local Better Business Bureau, bankruptcy court, and references.

NINE RULES FOR PLUMBING PROJECTS

1. Never run uninsulated pipes near outer walls, for when they freeze they create a considerable mess.
2. Inspect all grout and caulk with a bright flashlight and magnifying glass if necessary. Repair even the most microscopic holes in the grout or caulk. Make sure that everything that's supposed to be grouted, like the areas around the shower head and soap dishes, is grouted.
3. Assume that eventually all fixtures will leak in the worst way. Therefore, grout and caulk all floor surfaces in bathrooms.
4. Large tubs and spas are great ways to relieve the aching muscles you'll get regrouting things, but when the spa is installed, make certain that the weight is distributed evenly. That means constructing a platform for the spa or fastening

the side of the spa to the wall so that the wall carries some of the tub's weight.

5. Naturally, it goes without saying that the connections between all new pipes should be perfect and double-checked with a pressure test. Make sure that your plumber has double-checked.

6. Run all new appliances while the plumber is still around. This means starting the new dishwasher with dirty dishes. It also means taking a real shower. Seriously, how else will you know if the shower door isn't on correctly or that the shower water isn't directly connected to the toilet?

7. For the first couple of months (years, if you can remember), pay close attention to the ceiling below where the new bathroom was installed. If you notice any discolored paint, especially if it's mixed with cracking paint, call the plumber back at once. Most people don't look at their ceilings and, as a consequence, don't realize that there's a problem until real damage has set in. Houses are resilient creatures, and if you can catch a leak in the early stages, you're in luck. Ceiling watching was not my favorite activity after we moved into our newly renovated house, but then I realized that while doing my situps I could accomplish two important tasks at the same time.

Lest you think I'm belaboring this point, and should you be singing a chorus of "Get on to the next proposition already," let me tell you what happened to my brother-in-law, Richard, and his wife, Joanna, one summer afternoon. Proud of their new bathroom with dual shower heads, they spent considerable time there. Directly below was their living room, which featured many interesting objects at eye level but none in an upward direction. Richard and Joanna never looked up when they were in the living room and never noticed any of the telltale signs. Then, that fateful afternoon, they discovered part of their living room ceiling resting comfortably on their couch. By then, when looking up you could see from the living room into the bathroom.

Fortunately for them, the whole shebang was under warranty. Unfortunately, the company that built their bathroom wanted to send back the same plumber who originally had helped out with the union of their bathroom and living room. Richard and Joanna put a stop to that idea, however, by insisting that their contractors use another plumber.

8. Which brings me to point number eight. If you don't like the plumber, fire him. In other words, if the vibes are wrong, get a plumber you can have confidence in.

9. This may be your one and only opportunity to get the lead out. Seek out and replace any lead-soldered pipes that are in your house. If you're having a new sink put in and there's a now-visible pipe that leads to that sink, ask the plumber if it has lead solder. Replace that pipe, even if it wasn't part of your original plan.

Outside Water

Outside water can damage a house just as quickly as inside water. While the crews are diligently working on the various exterior walls, frames, and foundations of the ground around your house, moving earth, stones, and anything else in their way, make sure they don't alter drainage patterns. Water must flow away from the house. There should be a slope of at least one inch per foot for a total of twelve inches of slope in all. If you don't see this, your house is in danger of flooding—possibly even when you water the lawn.

The roof keeps water out of the top floor of your house, a good thing because most houses' top floors are not equipped to keep water out of the floors beneath them.

If anyone has been on your roof, there's a possibility that damage was done; particularly if you have a slate roof, because slates are very fragile. Never let a worker other than a professional roofer walk on a slate roof. It will be one of the most costly mistakes you can make. Like other areas of the house, the roof is most vulnerable where different materials meet. *Flashing*—thin strips of copper (better) or aluminum (more common)—is applied to angles, curves, corners, and other parts of the roof where the principal roof material doesn't fit tightly. Skylights, chimneys, and dormers are places where you're likely to see flashing. Flashing easily can become unattached, especially if someone on a ladder was hanging onto a part of the roof to balance himself. Be wary of this.

Looking in All the Right Places

There was a homeowner in Potomac, Maryland, who wanted to add a deck to his half-million-dollar house. The builder sent a foreman there, showed the foreman the plans, and said, "Build it." The deck was built two feet off the ground, as per the permits. The homeowner went out and loved his deck so much that he decided to throw a party. With fifteen people on board, the deck decided that it could no longer stay attached to the house and collapsed, much to the surprise of the partygoers.

The problem was that the deck was not bolted to the house. County code required half-inch bolts six inches long to be placed every four feet along the deck where it met the house, which would have allowed about fifty people on the deck. The deck was inspected, but the city inspector didn't catch that there were no bolts. Where the bolts were to go was an out-of-the-way-place and the inspector didn't look.

No one likes to get on a roof. Always check every square inch of work regardless of difficulty of access. If the roofer has gotten mad at you or decides not to put on a piece of flashing, you have no way of knowing if the flashing has been finished. The damage may appear later, but you have no way of knowing whether there's going to be damage until long after the worker has finished.

If revenge is sweet, then tradesmen are the sweetest people in the world. Revenge—against the contractor, against another worker, against you—is very common during home remodeling.

A siding contractor in Connecticut was angry with a homeowner because of personality conflicts, so the siding contractor put the siding on the house, but as he did so he made sure in one spot to nail the siding through the PVC plumbing stack rather than through the two-by-four. It took eight years for the plumbing leak to rot out a sufficiently large part of the wall to become noticeable to the homeowner. All the homeowner could do was rebuild his exterior load-bearing wall. It was only when he took off the siding to look at the water damage that he noticed the nails through the plumbing. Alas, eight years later the contractor was nowhere to be found, and the homeowner had to pay.

Don't Piss Off the Workers (Excuse the Language, but That's the Way Tradesmen Talk)

A few more words about not angering the workers, because if these guys can do damage when they're trying to do good, imagine what they can do when they are trying to damage your house:

In one New York State house a mason was constructing a brick veneer. This veneer was not directly attached to the house because wood and brick expand and contract at different rates. If solidly attached, the brick veneer will eventually crack. The veneer, however, was supposed to be attached to the house with metal strips at regular intervals to hold it to the wall. About one-half inch to an inch is left between the stud wall and the brick veneer; the metal strip is nailed to the studs and mortared in place. Okay so far? The homeowner and mason had a large argument about money, and the mason gave the homeowner his money's worth by not installing any strips. Later, the wall had to be torn apart and rebuilt at a cost of $1,500. All for a lack of five dollars' worth of materials.

A good manager will allow workers to do the job required in a way that makes the worker think that he is doing it his way. Giving tradesmen a sense of autonomy, while at the same time getting the manager the results he wants, helps ensure that workers maintain a healthy attitude toward your house. But this is a very difficult environment to create because not every tradesman works in the same way. Some workers want all their own authority, and others don't want any authority, refusing to do it any way except your way, even if you are wrong. A good manager walks the line between these two extremes and knows which workers want to be pampered. A good contractor knows the capabilities and personalities of his workers. Are they knowledgeable? Are they willing to give input into the project? Are they lackeys, followers? Preventing catastrophe is very much a component of keeping good rapport and communication.

You can force anyone to do anything under the contract when

you are witholding money until completed. If you are going to play the heavy, your ego may be satisfied, but your renovation could be a disaster.

Here are some more examples of conscious (or subconscious—it's hard to tell which) sabotage: Concrete can be poured very quickly in a foundation that doesn't have any steel reinforcement or plastic vapor barrier. You come back from work and voilà, the slab is complete. It'll take years for you to realize that the reason your concrete has deteriorated so quickly is that you (or the G.C.) acted like a jerk to the concrete pourer, who wanted to get out of your house as quickly as he could.

A roofer installing a cedar-shake (*shakes* are like shingles) roof in Tennessee didn't put the tar paper (called *felt*) layers at the right intervals between the cedar shakes. Roofers are supposed to put felt underneath the shakes to catch water that can drip through the cracks between the shakes. The Tennessee roofer pulled the felt down so that it would be exposed to sunlight between the shakes causing it to deteriorate and leak in roughly five years. Why? Because there were "payment problems" between the owner and contractor.

If you think sabotage occurs only among nations and corporate competitors, you've never been involved in renovation. When painters use a live-in carpenter's clothes to clean up their mess (because the carpenter told them to clean up in a not-so-pleasant voice), you have to be prepared for anything. Not long after a New Jersey couple moved into their condominium, their living room ceiling fell in. "A picture window in the living room ceiling" is the way the contractor who was called in to fix the mess described it. As repairs progressed, the contractor found a screwdriver in the water line, an obvious case of sabotage. But the problem was, after the fact no one could say who did it.

If you notice friction developing between any workers, intervene. It's far more likely that problems will occur among subcontractors than between you and the contractor because they see a lot more of each other than they see of you. Just because everyone working on your house may act like a good old boy doesn't prove that they are. While erecting your drywall, the drywaller actually may be fantisizing about a new shotgun and how it will make nice holes in the carpenter's truck. If you suspect a personality conflict, ask the parties if there's a problem. Separate the schedules of the afflicted individuals. If you have to, fire one of the antagonists. Keep an eye out for these kinds of problems because they happen frequently, and when one subcontractor doesn't like another, he may let that subcontractor know by injuring his work—your house.

Renovation sabotage is a very hard crime to prove—and workers know this.

Warning! Don't Rush the Important Things!

There will be times when you have to remodel quickly, but you should limit those occasions as much as possible.

Let me tell you a brief tale about the renovation of a $500,000 Delaware house. The homeowner was in such a rush to get the brickwork up that he didn't give the mason time to tarpaper between the brick veneer and the sheathing (a crucial board on the outside of the framing of a structure holding the studs square). He told the mason to move as fast as possible, no holds barred. This mason believed in following the instructions of the person paying his salary. The mason, like many contractors, did what the boss wanted, even though he knew it wasn't right. (Then again, there are always alternative methods of doing something.)

Masonry can hold up to 30 percent of its weight in water, and dry plywood behind it acts like a sponge to take moisture in. Over the years, the water jumped the gap between the brick and the the plywood frame of the house. The sheathing of the house rotted in ten years.

A Healthy Home

These days, environmentally safe and accidentproof houses are in. Lead-free paint, low radon levels, fire escape routes, nontoxic pesticides (including nonchemical termite killers), and antiscalding hot water valves are in. There are plenty of devices you can incorporate into your home to make it a safer place, from burglar alarms to nonslip surfaces on staircases. I encourage you to seek out these devices. These technologies and strategies are a little off the subject of this book; however, what is part of the subject is what contractors can do to make your house an environmentally unsafe place.

Lead paint is banned as an indoor paint. Too many children have gotten lead poisoning from eating paint chips. But lead-based paints are still used to paint the exteriors of houses. This might sound like no problem, but children do still eat lead paint from the outsides of houses. Worse still, lead paint is supposed to "dust off" to create a shimmer. In other words, in close proximity to lead paint, there is a high atmospheric lead concentration. Instruct painters not to use lead paint outside, and then check yourself to make sure that they do this.

Many houses built decades ago used asbestos as an insulator in a variety of places, including ducts, between walls, and around furnaces. The rule for asbestos is to leave it alone unless you are renovating. If you are reading this book, leaving it alone may not be an option.

Even if you are renovating, it's best to leave alone any asbestos that's not in the afflicted areas of your house. Only deal with asbestos that is being disturbed.

Always employ a licensed asbestos removal company to tackle an asbestos problem. In many instances the asbestos team will recommend *sealing* the asbestos rather than removing it, and usually that's the superior course. Asbestos only becomes dangerous when it's disturbed, so encapsulating it keeps the asbestos from becoming disturbed. Should removal be the method recommended for you, check to ensure that two things are done:

First, insist that the asbestos area be sealed off with a vapor barrier from the rest of the house while the workers are working on the asbestos.

Second, when the workers set up an asbestos monitor to determine whether there are any fibers left in the air, make certain that the monitor is set up inside. One McLean, Virginia, contractor told me he walked into an operation where "the testing system they were using to check the fibers was outside. It gave completely erroneous readings."

The Guest Syndrome

The guest syndrome has two common variants. You invite some friends to your house for dinner. While there, you present them with your ten-month-old baby, who has just learned his first words. What happens when the guests are an audience, waiting for your baby to talk? The baby remains silent.

Here's my version: You have a new window cut in your living room wall. During that work, the foreman notices that there is some rotten wood in the wall near the window. You replace that wood. Everything is fine. Two months later—after the workers have gone—the wall collapses.

If you notice an unexpected problem during renovation, assume that there is more to the problem than you can see. Deal with that problem while the workers are on the job.

The older the house and the longer the period between renovations, the greater the chance that you will find something structurally amiss. A North Carolina family who lived in a brick colonial was converting an old screened porch into a den. When the workers got to the stage of repainting the outside railing, they found that there was practically no railing to be found—the paint was all that was holding up the railing. The couple, naturally, wanted to replace the railing—a good idea—but the contractor said that they should replace everything connected to the railing. The fact that there was a water-damaged railing was clear warning to him that water could have damaged other parts connected to the railing. By replacing only the posts at that time, the couple would have acquired a constant damage-control (maintenance) schedule for the future. As it turned out, when the contractor knocked into the lower support wall, there was virtually no interior stud wall left. The exterior siding and interior paneling to the room were all that were holding up the porch roof. Dry rot and carpenter ants had destroyed the inside. Replacing the studs, posts, and wall (from the outside, so that the interior wood paneling wouldn't be harmed) cost about $1,500. Had the couple waited until the wall started to collapse, the work would have cost about $20,000.

Foundation, Footing, and Framing (Or, Making Sure That the Basics Are Right)

FOUNDATION WALLS

There are two types of foundation walls, poured concrete or cinderblock. When you pour a concrete foundation, the pourer must set up forms—panels that hold the concrete in shape—and put in rebar, if needed. During this process you have to watch slump. Patience is an important component of foundations. Don't let anyone *backfill* (do the initial grading around the foundation) until the concrete has completely cured, which will take about two weeks, depending on the temperature, humidity, and slump of the concrete. If you backfill too soon, you will pop the wall—that is, get a horizontal or vertical crack through the foundation where the concrete wall will bow. (If this happens, you need to reengineer the wall with new supports.) Concrete allows you to backfill before the rest of the house is constructed, just as long as the cement has cured. With cinderblock, you must wait to backfill until the rest of the house is constructed.

During the backfilling, it's crucial to keep machines—backloaders for example—six to eight feet away from cinderblock walls or they will pop, an undesirable condition. Sometimes workers reinforce cinderblock with rebar and pour pea gravel cement into the cells, the sign of a conscientious team.

FRAMING

A leaky toilet. A ceiling light fixture that falls out. A central air-conditioning system that doesn't cool too well. These are unpleasant renovation outcomes, but none in and of themselves are disasters. There are plenty of places where a mistake can lead to catastrophe, but one area more than others stands out as a place where precise and careful work is necessary: framing. If your bathroom ceiling collapses into your living room, that's a disaster, with a several-thousand-dollar repair bill. But if the framing for your house was done improperly, the only way you can repair that is to tear the house down.

This information is particularly useful if you are planning to be your own general contractor.

Everything done outside should be plumb, square, and level. Anything not so must be done all over again. (See the sections on framing carpenters and masons in Chapter 9.) What follows is a short course on what to look out for during this crucial stage of construction. The next few paragraphs aren't all you need to know, but they will provide guidance that can help you stave off the most serious mistakes.

When you dig footings for your addition or your house, generally the concrete will be about two feet wide, one foot deep, and run around the perimeter of your construction. The excavated ditch should have square corners at the bottom, and the bottom should be below your local frost line, about thirty inches below the Mason-Dixon line (in Pittsburgh it's four feet!). Don't accept any other geometrical configuration, and don't accept a too-shallow ditch.

Masonry products cannot be laid down faster than the mortar dries. If mortar is wet, then as you put additional weight on it the mortar will slip out and you will get a sagging foundation. If, during construction, the brick or cinderblock gets rained on, that's considered wet. Don't let the mason use either until they are dry, no matter which relative's wedding he has to go to.

Framing makes the house or addition begin to look like a structure. The important pieces of wood are called studs and joists, and they are weight supporting.

All lumber has twists, checks, bows. The carpenter needs to remove deformed lumber from the usable pile, otherwise your frame will be deformed. If you see your carpenter using all wood, assume that he's using deformed wood too and make a careful inspection of the material he's putting into your house.

When the carpenter spans a space with joists, you want to make sure that the lumber picked doesn't create a bounce in the floor. You don't want to feel the floor bounce, and the larger the joist, the less the bounce. If your house calls for prefabricated trusses, leaving you with large spaces with no support walls, be prepared: You are going to get a lot of bounce.

ROOF

In one whole subdivision in Northwest Washington, D.C., the houses had slate roofs with fine grades of slate on the front of the house and a lower grade of slate on the back of the house. The term for this is cheating. Look for this when your roof is being done.

Another form of cheating—saving money for a builder—is to use galvanized nails instead of hot-dipped nails for a slate roof. Galvanized nails get scratched on the surface when the slate expands and contracts over the seasons. Humidity rusts the nail heads off, and the slate may fall off. A *hot-dipped nail* has a zinc coating that resists rust from scratches. If you're using slate at hundreds of dollars per square, get the right nails, which cost just pennies a nail. Galvanized nails should be used on asphalt (fiberglass) and cedar-shake roofs, as well as on aluminum flashing. Copper nails are used with copper flashing.

The Typical Renovation

There is, of course, no such thing as a typical renovation. However, many of the stages that a given home renovation goes through are the same. And what is the same in many instances is the order in which work should be done. What follows is a description of the process that building a two-story addition should follow. If you are doing this kind of remodeling, you're in luck—just read on. Make sure that nothing mentioned in this section gets skipped on your project. If the work you are doing incorporates any of the elements of a two-story addition, and most renovations have at least some of these elements, then pay attention to those parts.

I've skipped a lot of detail in describing this typical renovation, but none of what is in this section should be skipped in your house. If your contractor skips any of these stages, ask why and insist that the work be done anyway.

INSPECT THE PROGRESS SO FAR

The typical construction job follows a general pattern of design and construction. This goes for practically anything—constructing an addition or other extra space. In the design process you solve all your problems of dimension and location. Once you have determined that you want to go ahead and execute your design, the builder will excavate the area where the expansion is to occur. This excavation is supposed to provide a level area the size of your addition on which you can build the platforms that are going to hold the remainder of your construction.

INSPECT THE PROGRESS SO FAR

The first and foremost item that gets built is called the *footing.* This is concrete poured in a trench around the outside perimeter of the construction area that distributes the weight of your addition onto a larger soil area and keeps your addition from sinking. The proper terminology is a *spread footer.* The basic rule on the footing is that it must be below the frost line for your area, as frost has a tendency to heave dirt, rocks, and concrete up in the winter and allows them to fall in warmer temperatures—not good for houses.

On top of the spread footer go the *foundation walls.* These foundation walls rise from the concrete and form the perimeter of the addition. They can be made out of cinderblock or poured concrete. The foundation wall generally rises up to the level of where you want to construct the first *deck,* the framing for your first floor.

Let's assume that your addition is going to have a basement room, and that the basement room is going to have plumbing installed. What happens is that the space inside your new addition gets filled with gravel to the tops of your footers—generally four to six inches above ground. The plumber then is allowed to come in and run what are called *ground works.* Ground works are the drainage systems that need to be below the house to tie in with the sewer. In addition, if you are in a waterlogged or flood sensitive area, you may want to install an *interior drain-tile system,* in which a perforated pipe is run around the perimeter of the foundation at the footing level and drains into a *crock pot* typically in a corner and used for installation of a sump pump.

INSPECT THE PROGRESS SO FAR

Only at this point can you think of pouring your *concrete slab* (the floor). However, this concrete slab needs attention first. The builder must lay down a sheet of plastic called a *vapor barrier* along with wire mesh. These aid in the curing and maintenance of strong concrete.

On the outside base of this foundation wall the builder should lay perforated drain tile that flows to an open swale area or ties to the inside sump system. This drain tile is covered with gravel and plastic. The foundation wall is then treated with damp-proofing material, ashpalt-based tar, and backfilled with dirt. If the weight of the materials used in construction or the weight of the dirt that will backfill the wall is excessive, as when using stone or brick veneer or trying to hold back a forty-foot hill, you need to add *rebar,* steel rods that reinforce the concrete—both for the footing and the foundation.

This completes the foundation and footing—the structure on which the addition rests.

On this foundation the first deck is laid. However, again you must prepare the foundation to accommodate the change in materials. (Remember, it is where different materials meet that your weakest spots occur.) The top of the foundation wall should have anchor bolts secured with washer and nuts and mortared in place, which will secure a *sill plate,* a pressure-treated piece of wood on which the joists will be laid. The contractor needs to put in a termite shield and sill insulation prior to installing the sill plate. Now your house is ready for the framing of the addition.

While performing the preceding tasks, constant attention needs to be given to the elevation of the foundation, so that the first deck will be level with the existing first floor of the house. With a perfect job you won't see the transition between the floors in the existing house and the addition. The converse happens too. The *joists* are the house's basic framework for flooring and are laid across the room onto the top of the foundation wall. If the series of joists abut the existing house, a plate must be bolted to the house and the joists secured with joist hangers to that plate. On top of this structure plywood decking using tongue-and-groove plywood must be secured with glue and nailed to the joists. The builder should screw it down (if you don't want the floor to squeak), unless you like squeaky floors. This deck is the house's base for constructing the exterior perimeter walls. The perimeter walls are constructed on this deck in *panels,* which consist of a bottom plate, stud walls, and top plates with all openings (windows and doors) in place. This framework is sheathed when square and cross-braced. The panel is then raised and nailed to the decking to form the outside perimeter walls. The builder must check that these walls are square, plumb, and online. By use of springboards the wall can be held in place for the next stage of construction.

The process of building a deck and then building panels can be repeated as many as three times to create three stories in platform-framing construction.

The last stage involves building the roof for this structure. The roof is supported by roof rafters, which slope from the outside walls to the peak. There they are held in place by a ridgepole. The roof is then sheathed with plywood (usually), tar-papered, and then shingled.

This provides the basic structure on which the windows and doors can be installed, which of course you ordered four weeks ago. The exterior treatment can now be constructed—siding, brick, whatever. The cornice work at the roof is completed and the gutters hung.

If the house's exterior finish is a brick or stone veneer, you need to upgrade from an eight- to a twelve-inch cinderblock foundation wall. This means that the deck and stud walls are set back four inches from the front face. This additional four inches is called the *brick ledge* and is what the masons will be building on.

The exterior is done!

Now on to the inside. The carpenter remains on the job until he has completed all the interior partitions, which are the walls for the rooms, closets, and so forth. The house is then turned over to the mechanical contractors. Generally the plumber comes first and does the rough-in plumbing,

INSPECT THE PROGRESS SO FAR

the mechanical contractor comes in and does the rough-in HVAC system, and then the electrician comes and does all the rough wiring.

INSPECT THE PROGRESS SO FAR

Rough-in is a term that describes the preparatory work that needs to be done within the walls or the floors. The mechanical contractors are responsible for proper placement of all fixtures that are to be attached to their systems such as sinks, chandeliers, switches, and heating vents. The insulator then comes in and insulates the outside walls using *batt insulation,* which is generally stapled to the studs, chinks, and open spaces around doors and windows and in the attic. He also fully insulates the attic and any other spots that need sound insulation.

INSPECT THE PROGRESS SO FAR

The drywaller then appears (hopefully) and covers all the walls with drywall, glues the drywall, and makes sure that he cuts openings for all doors, windows, switches, fixtures, and vents.

INSPECT THE PROGRESS SO FAR

The house is then ready to be trimmed. Trimming includes all finished carpentry, hanging interior doors, building shelves, and moldings. The tile can now be installed in the bathrooms (the tile you ordered two months ago) and the flooring put in. The kitchen cabinets can be installed.

This prepares you for painting your house. After painting, the final trim can be done: hanging light fixtures, installing plumbing fixtures and finished hardware (doorknobs, appliances, carpeting, and anything you don't want paint on). At this point, go through and make a big list of all the little items that are missing. It should take three to five days to take care of these items—a missing window lock or doorknob, caulking a window. This is called *punch-out.*

After completing punch-out, about 98 percent of the work is done.

INSPECT THE PROGRESS SO FAR

The remaining work concerns items that will become loose and need to be replaced, defects that become apparent and need to be repaired, and the like. But at this stage you should be able to fully enjoy your newly created area.

11 | Solving Postremodeling Problems

(Especially When You Don't Want
the Same Clown Who
Screwed Up in the First Place
to Fix Things)

NO RENOVATION IS FLAWLESS. If you know someone who hasn't complained about his renovation, if everything
seems blue-sky perfect to him, then he's either in desperate need
of a new eyeglass prescription or just lying. It's perfectly possible
to avoid postrenovation problems—by selling your house before
the remodeling is completed—but that's the only way. Humans
are far from perfect, and tradesmen are the most human of us all.

There are a number of areas that are especially prone to postrenovation complications. Pay particular attention to them, so
that a) you can catch them during the warranty period and b) you
can catch them before the problem turns into a disaster. (See
Chapter 10, "Preventing Catastrophe," for more on what you
should be doing during the renovation to keep your house safe.)

What follows is a list of problem areas on which you should
focus your eagle eye:

Roofs

Over time all roofs leak. Roofs that were poorly constructed leak earlier than those that were put on correctly. No
roof is constructed perfectly, and it is in those imperfect parts
that leaks will probably spring first. Small roof problems always
grow into bigger ones.

216

Skylights

Skylights beg water to come through them. If the skylight is new, there's no way to know whether it will hold up until the first torrential rain or until the snow on top melts into your house. If skylights are old, they will eventually leak when age catches up with them. You can't win with skylights (though they are wonderful). Be prepared to seal your skylights.

Incomplete Bathroom and Kitchen Caulking

Water will flow wherever there is a gap. All spaces around plumbing or water in your bathrooms and kitchen should be thoroughly caulked (but caulking should not be a substitute for properly placed fixtures). Tile should be well—read perfectly—grouted. Where tile meets a sink or bathtub caulk, not grout, should have been used. Expect that grout (and to a lesser degree caulk) will shrink over the first few months, so be prepared either to call the contractor or ask the contractor to leave you unopened grout and caulk so that you can do the inevitable.

Inadequate Drainage

Sometimes the only way to tell that drainage isn't good is to be surprised after a rain or after you've watered your lawn for the first time after renovation. I highly recommend a product call Water Bug, a small, battery-operated electronic sensor that sounds an alarm when your basement starts to get wet. Walk around your house: Unless you see the land clearly sloping down and away from the house, be prepared for some sort of drainage problem. Does the driveway slope away from your house? Backfill does settle, so something that's not a problem now may become one soon. When you see a small amount of water in your basement, don't say to yourself, "Let's see if this happens again next rain." It will.

Incomplete Exterior Work

Caulking and painting that's not finished, masonry work with gaps—these are typical postrenovation problems. Look for them especially around windows, siding joints, and where ducts and pipes enter the house. An unsightly omission now will become a leak later.

Substandard Structural Work

This is a tough one to fix if you notice it after the renovation is complete. A floor that bounces too much could be a sign of improperly seated floor joists, or joists that weren't made strong enough at the load points. A floor that sags indicates that this problem has become acute. In a new room or house you should not have to put up with an unlevel floor. An unlevel floor does not, as the contractor might want you to believe, result from unlevel ground but is probably a consequence of poor coordination among the tradesmen who were involved in building the addition. Unlevel floors can happen when sill plates are not flush with foundations which are not smooth and level. You could ignore this predicament (as fixing it can be a headache) and simply locate your baby grand piano elsewhere. But as with most problems, it is best to confront them immediately, especially if construction is still in progress. Houses, unlike the human body, are not self-correcting.

Poor Finishing

I read once that the major difference between a BMW and a Dodge is that engineering on the BMW is so precise that when the door is closed there is barely a measurable space between the door and the body of the car, and the space is perfectly even throughout. Badly hung doors, bad miter cuts, spackling that shows, shabby drywalling, weakly attached trim— all of these will manifest themselves later as bigger problems. Paint will fall off, walls will show cracks, air will enter homes, trim will wobble outward, doors will stick.

Old Systems Made Inoperative

Sometimes, in order to attach new electrical systems, the electrician needs to disconnect an existing outlet or switch. Sometimes these former outlets aren't necessary, and sometimes the decision the electrician made on his own can become a royal pain. A hot water heater may have to be disconnected to reroute the heating. You may also find that you no longer have access to old crawl spaces, and attic vents may be shut off. In addition to testing the new appliances, don't forget the old ones (an outlet tester is a good tool here, as is a working light bulb you can screw in).

Misaligned Exterior Windows, Doors, and Fireplaces

These let in a bit of cool air (plus noise, bugs, dust, and other unwelcome substances.) Doors, windows, and fireplaces are spots where different materials meet and are likely candidates for problems. You might have to wait until winter to get a sense of whether these barriers to the outside were put on correctly. If you notice any draft coming through the edges, call your contractor. Often caulk can be the savior, but sometimes the window or door has to be reinstalled.

Wires or Pipes That Go Nowhere

They may not be needed anymore, but on the other hand . . . any wire or pipe left dangling is a potential danger. There may be nothing at the business end, the end you can see, but you can't be sure that the other end is safe. Over time you're going to forget what that wire or pipe doesn't do, and it's going to cause trouble. Remove them.

When the Job Is Done . . .

The absolute first thing you should do when the work is completed is:

Change the locks.

The first thing you should do with every new appliance is:

Plug it in.

Really. I'm 100 percent serious about this. But I don't mean, If the TV doesn't work, plug that in. What I mean is, If something doesn't seem to work, connect the original valves—or whatever. Let me present an example to demonstrate what I'm talking about. When the renovation of our 1905 house was completed, everything looked grand. There was a lot of dust in the air still and a lot of dust all over the place, but as far as we could tell, it was all systems go.

Until we tried to dry our first batch of washed clothes. Sure the dryer tumbled around, so we knew it was working, but it didn't get hot. (Took us a while to figure that out too. Who suspects that a dryer that goes round and round doesn't get hot?) After we came to the conclusion that the dryer was broken, we called the company whose name was affixed to the front of the dryer with a little sticker. The sticker had yellowed, but the phone number, a combination of letters and numbers, still worked. They came. The fix-it men said the gas to the dryer was turned off. They charged us fifty dollars.

Apparently, for safety's sake, the workers had turned off all the gas throughout the house. Later the stove was reactivated, and so was the hot water heater. What normal person would have assumed that the gas to the dryer was turned off and never turned back on? Not us.

As this book is based on learning from other people's mistakes, here's one lesson you can learn from Peggy and me: Check to see that all connections are reestablished after the workers have gone. Then

Get instructions.

Every appliance should come with a manual. Every remote-control device should be accompanied by instructions. Take inventory now because it may not be possible to get instructions later. In addition, you should get a thorough explanation from your builder of how everything works. "Everything" includes how new-fangled windows open, how screens are installed, how the tub's faucets turn, how the HVAC system's filters are cleaned, how bulbs are changed in the fixtures. Don't forget to ask what's newly hooked up to the fuse box too.

Settling

It's done! The work is over and you can move back into your house or tear down the plastic that's been protecting the one sanctuary in your home, your bedroom. What a wonderful feeling! What a joyous day! What a reprieve for your checkbook!

And everything looks so beautiful (or is it just that you can't see the flaws because your glasses are permanently encased in plaster dust?). No matter. This is the day you've planned for a year.

I hate to be the bearer of bad news, but your contractor probably didn't tell you that things here and there are going to . . . well, for lack of a better word,* break. Switches are going to refuse to respond, no matter how much you press them; water is going to dribble here and there; some rooms will be cooler than others despite the new $10,000 HVAC system; windows will be stuck shut; paint and plaster are going to crack; hot water won't be; recessed lights will succumb to gravity; door knobs will want to stay attached to your hand, not the door; and whole appliances will go on strike, or strike out at you.

I hope you read this chapter after you've unequivocally decided to march ahead with your construction project.

When things go wrong, you will invariably call your contractor (who may or may not be on the vacation that you just paid for). Should your contractor be in, he will tell you not to worry, that after extensive renovation (or minor renovation, if your work was minor) every house goes through a process known as settling.

---

*A contractor-friend suggested that I use the expression "fail to perform fully," but that wouldn't have been completely accurate, would it?

222

A Helpful Hint

Discover your contractor's car phone number.
Whenever it looks like disaster is lurking beside
your dumpster or waiting to sneak into your
house, call the contractor on his car phone.
Explain the impending problem in incredible
detail. Cellular phone owners pay for the calls
they *receive*, to the tune of forty to sixty cents a
minute. At those rates, your builder will quickly
decide it's less costly to come over to your house
and see what's the matter than talk about it much
more.

How nice. *Settling.* Sounds like a passive, philosophical stage
that your house is going through. Settling. Such a pleasant word
couldn't bear any relation to the gigantic discolored crack that's
emerged directly below what you calculate is your new whirlpool
bath. Settling couldn't have any bearing on the total electrical
failure that pervades your new home.

Settling is that period of time in which the new things in your
house—appliances, walls, fixtures, etc.—adopt an antagonistic
attitude toward you. If one were to use a medical analogy, settling
would be to home renovation what tissue rejection is to organ
transplants.

Settling causes anger and more. If you resisted an ulcer during
the renovation itself, postrenovation disasters may still give you
one. You want nothing else than to enjoy your new home, but you
can't when everything seems to be falling apart and the parade
of workers continues. Some of these postrenovation problems
are preventable, especially if you adhered to the precautions
described in this book; others, no matter how skilled or how
expensive the workers, are inevitable. One of the frustrating aspects of settling is that while it's unavoidable, there's no way to
predict what your postrenovation particular problems will be.
Postrenovation disasters are what Murphy's Law was written to
describe. Anything that can go wrong will.

And, if I can add Adler's derivative of Murphy's Law: It will go
wrong when you are on vacation.

Not all postrenovation calamities are created equal. Settling problems fall into one of several categories:

EMERGENCIES

For example, a bathtub being deposited in the room below or a wall collapsing.

SOON-TO-BE EMERGENCIES

For example, grout that's shrunk inside the shower or a window that's loose.

THINGS THAT JUST WEREN'T DONE RIGHT

For example, a door latch that doesn't hold the door in place or a light switch that has to be jiggled to turn the light on.

THINGS THAT WEREN'T DONE RIGHT THAT COULD BE A REAL PAIN TO FIX

For example, a doorknob that was drilled too low (fixing it would mean leaving a noticeable surface where the old knob was) or the wrong window that was installed.

THINGS THAT ARE A LITTLE WRONG, WOULD BE A LOT OF TROUBLE TO FIX, BUT MAYBE YOU COULD LIVE WITH THEM

(Subset of the above category.) For example, a tiny crack in a window or wall-to-wall carpeting that pulls up a little at the edges.

THINGS THAT YOU AREN'T SURE ARE RIGHT OR WRONG

For example, the trash compactor chute that's seems harder to open than you suspect it should be or the refrigerator that juts out too far into the kitchen.

THINGS THAT FIT INTO ANY OF THE ABOVE CATEGORIES THAT AREN'T UNDER WARRANTY

For example, a little plastering job that you did yourself or a pipe that bursts 366 days after installation.

THINGS THAT ARE REALLY WRONG AND DOING SOMETHING ABOUT THEM MAY NOT BE WORTH THE TROUBLE

For example, a shower doorframe that isn't plumb, causing water to leak out of the bottom of the shower door or a heating duct that was placed in the only place where you can put a couch.

Each of these problems must be dealt with in proportion to how dangerous they are and how much they annoy you.

Testing After the Renovation

Other than seeing that the appliances run, the doors and windows open, and the heating and cooling system works, normal wear and tear is the only testing that you need to do. Don't put things through a major stress test by, for example, using your hallway as a bowling alley to test the strength of the finish. But use everything. This means that even if your renovation is completed in August, you should test the heating system.

The best method of testing construction is close visual inspection at intermittent points—once a month or every couple of months. Do look for problems! Fortunately (or unfortunately), malfunctions of machinery and mechanical systems will be very obvious, as will paint and plaster problems. More subtle things such as drywalls popping, floors buckling, carpets stretching, floor boards bowing are generally less noticeable, but in most cases these things are covered under builder warranties (assuming your builder is still in business).

If you spend an hour walking through the house after the renovation, you have a very good chance of picking up any signs of potential problems. When builder Paul Locher walks into anyone's house, he automatically thinks, Gee, this is a nice house. There are only twenty-seven problems. If you have gone through the construction process carefully and you've learned how construction methodology works, you will be more attuned to any vagaries at the end. If you look, you will see problems.

Are there any problems that should not be fixed? All problems should be fixed in time. Nothing in construction gets better on its own accord. Some components you want to allow to reach their maximum disrepair while still under warranty before calling in the troops because the longer you wait, the greater the chance that there will be multiple things to fix. A few weeks after your renovation there may be a couple nails that have popped from the wall, but after a couple of months, the whole sheet could be loose. On the other hand, maybe if you'd put the nails back in the sheet, the sheet wouldn't have come out. To decide, you have to think back to the basic construction methods. Did the guy glue the drywall to the wood? (Yes? That's good). Did he put screws in instead? (Yes? That's good.) If the construction was generally sound, the problems that occur will be slower in coming and less severe.

Go back to the basics of construction. New lumber settles after it takes a load. That is, it shifts, turns, and bends. How much, however, is the issue. A typical two-story construction with new white-pine lumber will compress the lumber about three-sixteenths of an inch. This invites minor drywall cracks at joints and nail pops in the drywall. These are minor in nature and don't mean that your house is about to collapse: these problems can be taken care of at your leisure or when the cracks in the paint drive you crazy, whichever comes first. Much more serious are vertical cracks that separate materials including cinderblock, brick, concrete, and drywall. Those cracks are signs of serious problems in construction and need to be addressed immediately. Immediately means *immediately.* An old rule of thumb is that a crack is not a serious crack unless it approaches one-fourth inch. However, even that rule has a caveat, because a quarter inch allows water infiltration into the house, which can cause other problems. So you have to modify this rule, depending on the rainfall in your area and where the crack is located. But this rule is still a good one because it tells you to look for the problem behind the problem, such as poor subsoil, cracked footing, or shifting loads. Any vertical crack that splits a material is a danger.

You will have cracks that split different materials or split two pieces of the same material in different directions (i. e., east-west and up-down cracks). These are less serious problems than one-way cracks within the same material. However, if these kinds of cracks appear in structural material, fix them! For example, a crack where a stud wall meets a brick wall isn't necessarily a very serious problem; neither is a crack in a piece of trim material, such as crown molding separating from the ceiling. One is more treatable than the other, however. All the caulk in the world will not keep a stud wall from separating from a brick wall, but it may keep your crown molding from falling off the ceiling.

If you do a minor cosmetic repair, like putting the trim back in place, continue to monitor the material for further separation. If the separation approaches the one-fourth inch, find out what the cause is.

For appliances and mechanicals, remember that everything has its wear factor. It isn't advisable to do engineering-style testing by turning on and off the valves fifty thousand times in the first month, because that will spur on the breaking. Be content that the fixture will eventually break, and you will know when that happens.

Hiding Mistakes

Every good contractor knows how to hide mistakes. This isn't necessarily an evil thing because you want the job to look good, and looking good includes putting in extra trim to cover the gap where the drywall didn't meet the window. Many mistakes can be hidden away behind wall—literally.

The drywaller is the one whose job it is to make minor errors invisible: walls that bow, utilities that protrude from the wall, framing out of square.

We had an oval window installed on the side of our house. Unfortunately, the window was tilted about five degrees off its axis. To compound the problem, the window was in our bedroom, so we were bound to see it every single day for the rest of our lives. Clearly, this was a "bad stress factor" that would only compound over the years, eventually causing us to defect to Albania. The contractor assured us, however, that plastering would alleviate this eyesore and that we'd hardly be able to notice the tilt. Indeed, re-plastering around the window somehow—and I can't explain how—made it look straight. In this case—though this is not true for everything in home construction—if you can't see a problem, it doesn't exist.

One couple wanted to place a mirror on the wall above the

sinks in their bathroom. Problem was, the wall was bowed, so the mirror stuck out three-quarters of an inch on the left side. But the tile installer put the tiles on in such a way that they wrapped around the mirror. The result? No gap between the wall, tiles, or mirror. As an added benefit, the wall now appeared straight! The wall is still bowed, but the bowing is only a theoretical problem. The solution that the tile installer offered was exceedingly important because this couple stands in front of the mirror every day.

The Final Checklist

Contractors call this the *punch list.* By this stage, you probably will feel that way yourself. The punch list is the list of things that have to be done before the job is considered complete. You may notice that your punch list is longer than your contractor's list. If the two of you can't agree on the substance of your respective lists, have your architect arbitrate (if you have one.) Or ask the contractor to specify why he believes that certain items on your list should be excluded from the punch list. Sometimes during this process the two parties come to a meeting of minds. It may be that you have to let certain minor things slip by—like painting a molding or getting the unused two-by-fours removed from your garage—to have the more important work completed.

The punch list is composed of everything you think needs to be finished or redone before the job is complete. Be thorough. Go through each room foot by foot, examining trim, framing, floors, windows, outlets (test them), fixtures (test them), plumbing (now's a good time to take your first shower in five months). Did you get screens for each window? Do all the doors lock shut? Are all the three-way switches hooked up correctly? Does your telephone wiring connect correctly? Does the paint reach into the hard-to-get corners of each room? And so on. Take a flashlight and tape measure with you.

Cleaning

If you've been cleaning as you've gone along, the final cleanup won't be too horrific. There are two ways to go about cleaning up. First, you can let your contractor do it, since your contract probably specifies that he will clean the renovation site after the work is done. Alternatively, you can tell the contractor that you'll do it and pay for it.

I recommend that you do it.

Why give the contractor these unpromised dollars? The contractor, in all likelihood, will hire the least expensive cleaning crew he can. A cleaning crew who thinks that removing part of a wooden surface along with the paint that is splattered on top of it is okay because what's important is that the paint is off. A cleaning crew that doesn't believe that walls are solid bangs your walls and chandeliers while carrying unused molding downstairs (okay as long as no one sees). A cleaning crew that washes the labels from new windows with glass cleaner, which pours onto your just-finished floors. A cleaning crew that takes debris out and tracks mud in.

Keep a Watchful Eye

Anything can go wrong with new construction. Be attentive to your new rooms and appliances. For example, the doorstop you have attached to a hinge might not be properly adjusted, and when the door is opened—Bam! A nice scar in your newly painted wall. Or listen to what happened to us: The cutoff switch on the automatic ice maker in our refrigerator wasn't attached properly. When we opened our freezer door after a week, it was like visiting Antarctica.

Maintenance

> Turning and turning in a widening gyre
> The center cannot hold . . .
> Things fall apart . . .

Was William Butler Yeats writing about home renovation? Almost every part of your house needs maintenance. Some maintenance tasks are so obvious that you don't even think about them as maintenance: painting every several years, cleaning the fireplace once or twice a season, replacing the bags in the garbage compactor every week.

Other systems require maintenance on a less predictable, or should I say less regular, schedule. Some new parts of your house may require you to maintain them in ways you never thought you would—or could. I'd like to introduce you to a couple of these systems. You could easily ignore these systems, but that might also be the downfall of your house.

AIR CLEANERS

Did you get one of those electrostatic air cleaners for your HVAC system? They're great devices, trapping particles only microns wide, reducing the flow of pollen, dust, asbestos fibers, bacteria, and other noxious matter through your house. I was introduced to electrostatic cleaners by my dentist, Dr. Steven Newman, and I am forever grateful because the pollution level in our house is significantly lower than the pollution level outside than it would normally be.

But there's a downside to these "April air fresheners," as one friend calls them. The filters have to be cleaned twice a year, and there are eight heavy metal filters (with sharp edges I discovered) to remove, soap, wash, rinse, dry, and put back in the correct sequence. Actually that's not so bad either. But here's the catch: For the first six months after you renovate your house, the air filters have to be cleaned once a month. That's because there's so much dust floating around the house. It's a chore—and something you should be aware of. But cleaning the filters is a whole lot better than breathing the harmful materials that houses generate.

THE ELECTRONIC THERMOSTAT

If you can program a VCR, you're going to be a success. The thermostats used to control HVAC systems are small computers that need to be reprogrammed on certain special occasions including prolonged power failures (assuming your thermostat has a battery back-up); when you go from heating to cooling (something you may do several times during the same week); and, finally, during my favorite holidays, the twice-annual setting of the clocks forward and backward. It's always possible to ignore the change from daylight saving time, but not reprogramming your thermostat means that when you awake at 7 A.M., the temperature in your house will still be set for 6 A.M. (a chillier setting). Even more upsetting is when the thermostat says 8:15 P.M. while the clocks in the rest of the house are in a different time zone. You might want to replace your programmable thermostat with the old-fashion dial kind.

Lawns

Visions of America covered coast to coast by a vast expanse of lawn are quickly made blurry by the sweat pouring down your face as you mow mow mow. New lawns require a lot of care. They need to be watered several times a week. They need to be mowed more gently than well-established lawns. Later you can hire a teenager, but in the beginning you have to take good care of the lawn.

A Compendium of Sage Wisdom on Remodeling

Now is not the time to give a tradesman starting out a chance to try his skills.

Don't show up in a suit and tie and tell your workers what to do.

Your first words to painters should be, "Wash your brushes only in the basement sink. Don't clean them anywhere else, including on the trees outside." Translate as needed to other languages.

Don't rely on notes to communicate with workers. Verbal instructions supplemented with graphic signs (such as a picture of a ladder inside the international "ban" symbol to let workers know they shouldn't put the ladder in your new bathtub) are better.

If your cat disappears during renovation, don't assume that it has run away. Even if you don't hear any meowing, the cat may have been trapped—and killed—somewhere in the walls or between floors.

Don't talk politics with eastern European carpenters.

Don't plan a dinner party for the board of trustees of your company for the evening after the renovation is scheduled to be completed.

231

A remodeling carpenter is different from a new house carpenter.

If the exterior of your house looks good, has historic value, or fits in with the neighborhood, don't change the exterior as you remodel the interior.

Always make sure that your foundation construction is level, plumb, and square, because everything else will rest on top of it.

Never have your house be the first in the city with a new anything.

Never assume that a worker will speak enough English to understand your—or the contractor's—instructions.

Never assume that your instructions will be conveyed to a subcontractor unless you give the instructions to him yourself.

Inspect foundations.

If you don't want something done a certain way, say so; you will have to live with your decisions for a long, long time.

Only licensed plumbers should do plumbing.

Once you've made your initial deposit, pay no one in advance. No exceptions.

Do as much renovation work at once as you can afford. You may never want to do this again.

Don't pay anyone a final payment until you are absolutely certain that you don't need him on the job any more and that all his subcontractors have been paid.

No initial deposits for more than 30 percent of the contract.

No full payment unless materials are actually delivered to the job site.

Violating physics is dangerous. Respect load-bearing walls.

Don't run gasoline-powered generators or compressors inside with the windows closed.

When outside work is being done, assume that the workers will change the path of water flow around your house away from the house toward the house.

Never rush. Especially when working. Don't let anyone on the job rush either.

Your project is going to cost more than you expected. Have at least 15 percent more than the estimated cost of the project readily available.

Even if you have renovated before, you should not expect your second renovation to be anything like the previous one.

Just because your architect communicates in learned technical and philosophical phrases and the builder speaks in profanities, don't assume that the builder knows any less than the architect about what's supposed to happen to your house.

When deciding whether to hire someone, references are better than résumés. Hiring without checking references is synonymous with disaster.

Check current references. If a contractor or subcontractor won't provide up-to-date references, it probably means that he is hiding something.

Workers don't care one iota about your personal possessions, the condition of your house, dust, debris, or anything else that contributes to the quality of your life.

The main thought running through workers' minds is, "Is it Miller time yet?" And for some, Miller time started before they arrived at your house.

Pay as little as you can up front. He who holds the money controls the project.

If you deal with subcontractors directly—by paying them or giving them parts of your house to repair—make sure you know where to find them. Post Office boxes don't count.

Don't threaten your contractor with a lawsuit. They hear it all the time and know that, in most cases, a lawyer is going to tell you it's going to cost more to sue than you could possibly get.

Be prepared to be at home more than usual during your renovation.

Getting through on the phone to an ace carpenter is more important than talking with the president of the United States.

Good work counts for more than anything else.

When in doubt about the quality of work, hire an expert to inspect—a structural engineer, an architect, a certified home inspector—or call in the city inspector.

Don't hire subcontractors who don't own cars. You will end up picking up supplies for them.

Watch for uniformity of materials. When a worker runs out of a particular color of something—say brick—he will continue the work using something similar and tell you either a) the color is the same, or b) the colors will become the same over time.

When work goes from bad to good, assume that it will go back to bad again.

If you are having trouble with a contractor, find other people the contractor is working for and compare notes.

Change the locks after the work is over.

There is an inherent sloppiness to workers. There is nothing you can do, no power on earth that can change that.

Quality of work varies by region of the country. If you've moved from one area to another the work you get may be better—or worse. Penny Moser, a writer and home-renovation veteran, says, "The apex of competence in the United States is where Iowa, Illinois, and Wisconsin come together, and its disintegrates in concentric circles with pockets of exceptions as it moves toward both coasts."

Home renovation is a service, not a product. It's the quality of the people who make the difference.

If you decide to build a deck around a tree, you are betting that the tree will outlast the deck.

If you are afraid to watch something (a wall knocked down, a whirlpool tub hoisted over the roof) because you are afraid of what might happen, that's exactly the time to watch.

There's a mythical worker to whom all workers point, who will make everything right in the end. When paint drips on the floor, the painter says that the floor guy will get that out. When varnish gets on the walls, the floor finisher says the painter will get that.

Most subcontractors don't return telephone calls. The only message you should leave is when they should show up at the job. They either just show up or don't.

Wherever there is caulk, make sure it is complete and tight, not brittle and broken.

The exterior of your house that faces the worst weather—usually the south and southwest—will probably require the most work.

If exterior work was done to your house, make sure all gutters have been replaced correctly.

No one other than a professional roofer should walk on a roof.

If you buy a fixture through a contractor, you have every right to see the invoice to learn what he paid for it. That way you can determine if the markup is fair. Let the G.C.—and subcontractors—know in advance that you want to look at the original invoices for all materials.

Ask your contractor if it's okay to use a new fixture or appliance before deciding to test it on your own. An incompletely hooked up dishwasher is going to make a real mess of your kitchen.

Other than a level, the second most important tool you have when inspecting the job at the end of every day is a notepad. Make careful notes on anything that looks wrong.

When in doubt over what kind of material to choose—for windows, doors, floors, cabinets—try to choose the material that is most compatible with what you already have.

Never sign any contract without first (a) reading it, (b) consulting an attorney, and (c) sleeping on it.

You may have to decide between getting the work done or getting it done the way you want. Getting it done is a stopgap measure that will make you happy for months; getting it done the way you want will make you happy for decades.

If you have a dog, (1) board it, (2) tie it to something secure all day long, or (3) print LOST DOG signs.

An apartment or house undergoing renovation is dangerous to children. Keep them out as much as possible until everything is completed.

Desiring design means you want an architect. Desiring just space means you don't.

You have much more protection against shoddy work and dishonesty if you use licensed contractors and subcontractors.

If a subcontractor starts off by perfoming shoddy work, assume that all subsequent work will be bad. Fire him.

Index

237